Doré Beach

The Responsible Conduct
of Research

Distribution:

VCH, P.O. Box 10 11 61, D-69451 Weinheim, Federal Republic of Germany

Switzerland: VCH, P.O. Box, CH-4020 Basel, Switzerland

United Kingdom and Ireland: VCH, 8 Wellington Court, Cambridge CB1 1HZ, United Kingdom

USA and Canada: VCH, 220 East 23rd Street, New York, NY 10010-4606, USA

Japan: VCH, Eikow Building, 10-9 Hongo 1-chome, Bunkyo-ku, Tokyo 113, Japan

ISBN 3-527-29333-7

Doré Beach

The Responsible Conduct of Research

Weinheim · New York
Basel · Cambridge · Tokyo

Doré Beach
University of South Florida
Office of Research FAO-126
4202 East Fowler Avenue
Tampa, FL 33620-7900
USA

Published jointly by
VCH Verlagsgesellschaft mbH, Weinheim (Federal Republic of Germany)
VCH Publishers Inc., New York, NY (USA)

Editorial Director: Dr. Michael Bär
Production Manager: Dipl.-Wirt.-Ing. (FH) Bernd Riedel

Library of Congress Card No. applied for.

A catalogue record for this book is available from the British Library.

Deutsche Bibliothek Cataloguing-in-Publication Data:

Beach, Doré
The responsible conduct of research / Doré Beach.
– Weinheim ; New York ; Basel ; Cambridge ; Tokyo : VCH, 1996
   ISBN 3-527-29333-7

Composition: Graphik & Text Studio Zettlmeier, D-93164 Laaber-Waldetzenberg
Printing: Strauss Offsetdruck GmbH, D-69509 Mörlenbach
Bookbinding: Wilh. Osswald & Co., D-67433 Neustadt

Printed in the Federal Republic of Germany

No man is an island, entire of itself; every man is a piece of the continent, a part of the main . . . any man's death diminishes me, because I am involved in mankind; and therefore never send to know for whom the bell tolls; it tolls for thee.

John Donne, Devotions Upon Emergent Occasions

Table of Contents

Preface

THE purpose of this book is to introduce pre- and postdoctoral students planning careers in scientific research to ethics as applied to the particular issues that will affect their present and future practices as scientists. Although it may seem contrary to common sense that any researcher should need instruction in responsible conduct, it is also important to keep in mind that though many scientists may defend science itself as being ethically neutral, its study and pursuit are not value- free. In many circumstances several morally permissible options may appear; in others, no decision seems wholly free of negative consequences. It is important to remember that the scientist no longer exists in a laboratory isolated from the larger society.

Recognizing the increasingly complex impact scientific research is having on society, the National Institutes of Health (NIH) has required that all applications for training grants include a description of a program to provide instruction in the responsible conduct of research. Although the NIH will not establish specific curriculum or format requirements, all programs are strongly encouraged to consider instruction in the following areas: conflict of interest, responsible authorship, policies regarding the use of human and animal subjects, and data management. As these NIH requirements illustrate, the tradition of protecting researchers is giving way to increasing encroachment of external regulations, mandates, and policies. Therefore it is incumbent on us all to protect freedom of conduct through an educational process while we still have a choice. This book is written in the spirit of that choice.

The text begins with an introduction to the process of ethical decision-making and moral reasoning, and concludes with the manner in which the responsible conduct of research is transferred for the benefit of society. The assumption is that the student has had no prior experience with the subject, and, with that in mind, each chapter contains questions for discussion and study and recommended readings. Pertinent ethical dilemmas and case studies are also included where appropriate. The contents proceed from general issues of responsible conduct to specific policies, mandates, and procedures. The text is organized to enhance class discussion and interaction with guest speakers or experts in the field.

Presently, we do not know the precise relationship between human knowledge and human behavior. However, educators *do* know that appropriate role models, mentors, and significant others have a strong influence on the moral development of individuals, and that this influence can affect behavior. The intent of this book is to provide students not only with the knowledge and skills to make appropriate ethical decisions, but also the awareness that fosters integrity and responsible conduct.

Dr. Doré Beach, Assistant Professor
Director, Responsible Conduct of
Research and Applied Ethics
Public Service Programs
University of South Florida

Acknowledgments

IN the seventeenth century John Donne in his immortal poem recognized that "no man is an island, entire of itself." Today, perhaps more than ever before, scientists are beginning to understand the interconnection of life on this planet as expressed in Donne's poem. As the Human Genome Project proceeds, and biotechnology advances, there is the realization that the survival of all species depends on this knowledge and an ethic that supports this recognition. This book is dedicated to future research scientists who, one hopes, will never "send to know for whom the bell tolls."

The individuals who contributed to this text share with the scientific community an awareness of species interdependence, and for that I am grateful. They are Dr. George Newkome, Vice President for Research, University of South Florida; Bryan Burgess, J.D., Associate Vice President for Health Sciences Legal/Institutional Affairs, University of South Florida; Kenneth Preston, Jr., J.D., Assistant Vice President, Office of Research, Director of Patents and Licensing and Research Foundation, University of South Florida; Lawrence Oremland, J.D., Associate Director, Patents and Licensing, Office of Research, University of South Florida; and Dr. Richard Streeter, Director, University of South Florida Division of Sponsored Research. I wish to thank them for their fine sense of responsibility and commitment to this project. I want also to acknowledge the assistance of Tom Ferguson, Dennis Freeman, and Ron Larson of Compliance Services and the support of Drs. Lynn Wecker, David Morgan, and Joseph Krzanowski of the Pharmacology and Therapeutics Department of the University of South Florida College of Medicine, who understand the need for instruction in the responsible conduct of research. I owe special thanks to my editorial assistant, Cathy Anderson, for her guidance, and to my family and colleagues, whose enthusiasm and encouragement kept me on course.

Dr. Doré Beach

Introduction

Ethics is in origin the art of recommending to others the
sacrifices required for cooperation with oneself.

Bertrand Russel

THE framework of ethical wisdom is constantly challenged by new
problems arising from scientific and technical discoveries and inventions. These advances have created unforeseen and unprecedented
issues that require decisions often beyond the scope of our present-day
moral reasoning. For example, the traditional thinking, rules, and regulations regarding the termination of human life do not provide answers to
the question of how long life should be sustained by means of medical
technology. Because of medical advances, the debates over the definition
of death and the amount of effort that should be applied to sustain life are
no longer simply provocative academic exercises.

In the fields of genetic and environmental engineering, nuclear science,
and computer technology, ethical decisions are at the very heart of an array
of legal, social, and policy issues. These decisions profoundly influence
institutional practices and everyday life.

Physician-novelist Michael Crichton, in the introduction to his novel
Jurassic Park, makes the point that biotechnology research is proceeding
without regulation, and that there are very few research institutions that
do not have commercial affiliations. In the past, it was possible for researchers to separate themselves from the commerce of day-to-day life.
Today, however, research is considered big business. It is no longer, for the
most part, conducted only for the pursuit of new knowledge. In many cases
the research paradigm has shifted to a product orientation—the transfer of
that new knowledge to the industrial platform in order to reach the public.
Meeting the needs of society is a noble and well-intentioned result of
research, but the factor of commercialism adds a dimension to the responsible conduct of research. If scientists ignore the broader responsibilities
their endeavors entail, as in *Jurassic Park*, the dinosaurs may return, bringing
with them dilemmas far greater than history has recorded.

The advent of the 1990s brought the previously elusive topic of research ethics to the forefront. In the form of federal mandates, the National Institutes of Health (NIH) and the National Science Foundation (NSF) initiated a process that requires universities to have reporting mechanisms and inquiry processes that govern misconduct in research; otherwise those institutions would lose current and potential sponsored research funding. Institutions were faced with the complicated task of compiling documents amenable to faculty, staff, and administrators, and in compliance with the federal requirements. These documents were intended to resolve allegations of misconduct without impugning the reputations of the researchers when such allegations were determined to be unwarranted or unfounded, to ensure the integrity of the research and inquiry processes, and to reassure the funding agencies that universities are taking appropriate action to comply with the stated regulations.

Over the years, in similar moves to bring attention to ethical accountability, federal agencies have mandated additional measures related to the responsible conduct of research. Research institutions are required to have assurances on file with government agencies delineating the processes that will ensure responsible conduct in the use of human and animal subjects in research. Research involving radioactive materials is strictly controlled and monitored by government licensing processes, as is the handling of biohazardous waste and controlled substances. The drug-free workplace, Affirmative Action, and even government lobbying have forced ethical regulations on the once-independent researcher.

This past year NSF and NIH began the process of requiring research institutions to put into place policies and procedures to govern conflict of interest. Though painful in their bureaucracy, the government agencies, in response to public interest and pressure, have issued research university administrators and faculty researchers a wake-up call. Researchers must now adhere to numerous compliance mandates, ethical issues must be addressed in their work, and this work ethic must be brought to the classroom and laboratory. Learning to conduct research responsibly and ethically is a necessary and integral part of the education of today's scientists and scholars. Given the conflict-of-interest issues confronting research and the unknown factors the future may bring, a practical, hands-on compilation of the pivotal ethical issues facing faculty and student researchers is needed. It is our expectation that this text will fill that need.

Dr. George R. Newkome

CHAPTER ONE

Ethical Decisions and Moral Reasoning

> Moral wisdom is exercised not by those who stick by a single principle come what may, absolutely and without exception, but rather by those who understand that, in the long, no principle–however absolute–can avoid running up against another equally absolute principle; and by those who have the experience and discrimination needed to balance conflicting considerations in the most humane way.
>
> *Stephen Toulmin, The Tyranny of Principles*

Introduction

T HE GOAL of this chapter is to provide a guide for those planning careers in scientific research and a framework for making ethical decisions, as well as to stimulate thinking about ethical issues and the complexity of such issues as a result of the rapid developments in science and technology. The chapters following will focus on the concerns, issues, and ethical dilemmas associated with these new developments. This is not meant to imply that ethical issues are entities apart from the larger society, nor does it mean that scientific researchers have an ethic apart from their responsibilities to society.

The ethics discussed in this text are referred to as *applied ethics*, defined as the application of moral principles and codes to concrete rather than abstract conditions, especially those conditions in which the principles and codes make *conflicting claims* on the condition or situation under examination. When such conflicting claims occur, it is referred to as an *ethical dilemma*.

It is important to keep in mind that learning about the process of ethical deliberation does not directly influence a person's behavior. This process will not create a person who has more integrity. However, studies on ethics education do suggest that ethical development is not complete or fixed at the time most students attend graduate school (Rest, p. 22). Learning about ethics does foster awareness, and it can reinforce the importance of actions that constitute appropriate behavior in the conduct of research. Informed participation enables responsibility. A background of knowledge in ethics aids our recognition of *choice*. It is the recognition that choice exists, and the understanding of personal accountability, that empowers us to act responsibly.

The Language of Ethics

Because moral reasoning concerns itself with laws, facts, values, rights, standards or codes, and principles, it is necessary first to establish an understanding of the language of ethics (see Glossary). *Ethics* is defined as the discipline related to what is good and bad or right and wrong behavior, including moral duty and obligation, values and beliefs, and the use of critical thinking about human problems. Ethics is a classification of *philosophy*. Philosophy is divided into four areas: *metaphysics*, the study of reality; *epistemology*, the study of the nature of human knowledge; *logic*, the study of the validity of argument; and *ethics*, the study of morality (Fig. 1). Ethics is further divided into the categories of *metaethics* or *theoretical ethics*, the study of meanings of ethical terms and the forms of ethical argument; *descriptive ethics*, the study of moral and ethical beliefs and customs of different cultures; *normative ethics*, the study of ethical principles that have been accepted as norms of right behavior; and *applied ethics*, the application of moral standards used in decision-making to concrete rather than abstract conditions (Fig. 2).

In many texts ethics and morality have distinct definitions. In this text the words ethics and morals are used synonymously. *Ethical reasoning* is a *process of analysis* to determine what is right or wrong—what is the correct or more responsible choice in a given situation. It is also an examination of our moral judgements, and an attempt to determine the grounds on which these judgements are based.

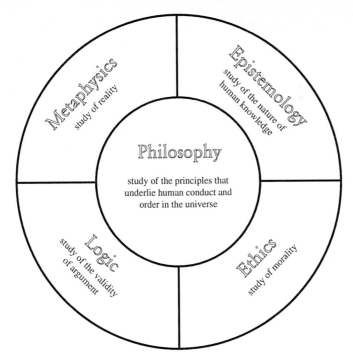

Figure 1 Philosophy

Ethical Theories

There are two types of ethical theories: those based on *principles* and those based on *virtues*.

Principle-based theories (normative ethics) are structured into two schools or theories. The first is known as *deontological*, the theory of *obligation* or duties, or rules and rights. The second is known as *consequentialist*, the theory that links the rightness of an act to the goodness of the state of affairs it brings about. This theory is also referred to as *utilitarianism*. In principle-based theories the judgements made may be general or specific. However, they are all normative in that they affirm or apply norms or standards to making decisions. To summarize: Principle-based theories (1) identify ethical principles, (2) evaluate ethical choices in terms of how well they fit with those principles, and (3) are abstract in the sense that the principles embraced must satisfy *universality*; that is, they must be applicable to all relevantly similar cases. They must also be *impartial*, which is to say, they must be objective.

Figure 2 Ethics: four related but distinct inquiries

Virtue-based theories are also structured into two schools: *communitarianism* and *relationalism*. Although virtue-based theories do employ fundamental ethical principles—for example, rules against lying, cheating or stealing—they rely first on the particulars of the concrete situation and then apply the principles. This process of deliberation is known as the *Aristotelian approach*, which means that "practical wisdom" is employed in the reasoning process, and the focus is on the uniqueness of each ethical situation. Communitarianism is based on shared community values or closed societies in which there are collective values shared by all. Relationalism emphasizes the values of love, family, and friendship inherent to the situation at hand. Both of these approaches take into account the unique features of an individual's personal history, affections, and family and community obligations. To summarize: Virtue-based theories (1) identify the ethically virtuous person, (2) evaluate ethical choices in terms of how well they exemplify the deliberations of the ethically virtuous person, and (3) are concrete and situation-based.

Imperatives

The theory that one employs in the process of analysis of ethical dilemmas depends on the nature of the question to be addressed as well as the role of the individual making the decision. All ethical theories capture perspectives that are important in ethical deliberation. Basic to all ethical theories are three *imperatives*, moral obligations or commands that are *unconditionally and universally binding*. These imperatives form the foundation and criteria for evaluation of every moral system authored by human beings. They are the source of every moral dilemma, but, though we can explain correct judgements in terms of these imperatives, we have difficulty deciding which takes priority over which. These three imperatives also form the foundation from which all laws, standards, and codes evolve. They form the core concepts around which the Declaration of Independence, the United States Constitution, and our system of laws are constructed. They are imbedded in the Judeo-Christian ethic through the Ten Commandments. The three imperatives from which values, moral duties, and choices emerge are:

1. *Human Welfare–Beneficence*: This relates to the idea of helping others—protecting them from harm, healing their illnesses, or saving their lives—and the duty to promote good, prevent harm (non-maleficence), and use the maximization of human happiness for the greatest number of individuals as the criterion for right action.

2. *Human Justice–Fairness*: This imperative requires one to set fairness for all above benefit for some. It encompasses the responsibility to apply fairness in all dealings with others, particularly fairness in the law, which is justice. Our form of law, distributive justice, requires that we seek the morally correct distribution of benefits and burdens in society.

3. *Human Dignity–Autonomy*: This relates to the idea of respect for persons, including their rights to choice, freedom, and privacy, and protection of those with diminished autonomy. The autonomy imperative requires us to respect the choices of others, and to allow them the space to live their lives the way they see fit.

Ethical Theory and Applied Ethics

At this point it is important to understand the relationship between the disciplines of theoretical and applied ethics. Many philosophers feel there is a serious gap between the two. Some of the criticism rests on the fact that within the principle-based theories there is disagreement regarding which of the two theories, consequentialist (utilitarian) and deontological, yields the *best* answers or solutions, and many philosophers have taken the position that virtue-based theories are not even theories. Given this division within the ranks, it is a wonder that there is any suitable approach a research scientist can use to solve ethical dilemmas. However, philosophy is not just for philosophers. We all use moral theories, albeit not always consciously, to justify our actions or judgements, and we make our decisions on a case-by-case or situation-by-situation basis. At this point one might ask, "then why bother to learn about how to apply ethics?" The answer to this question might be that we do not generally recognize or understand when there are conflicting claims on imperatives and values, nor do we understand how to justify the judgements made in the process of resolving ethical dilemmas (Fig. 3).

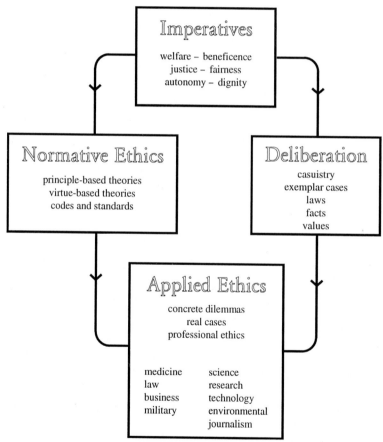

Figure 3 Ethics applied

Applied Ethics Today

In the early 1970s the discipline of *bioethics* emerged. Defined as the study of the rightness or wrongness of acts performed within the life sciences, bioethics applies both ethical theory and casuistry, commonly called the case-study method, to the resolution of moral dilemmas created from the complexities of contemporary developments in science and technology. The necessity of finding solutions for these complex dilemmas has forced the philosopher to move from the abstract and theoretically ideal situation to applying ethical reasoning to real situations that researchers in laboratories, hospitals, and society in general live with every day. In the future, new theories of ethics may emerge, and existing theories will be supplemented to accommodate the demand for reality-based solutions. The time has come when ethical theorists can no longer ignore the problems of application, and those working within applied ethics can no longer operate effectively without taking theoretical considerations into account. In addition, moral philosophy is becoming more intertwined with political, social, and legal philosophies. The notion of *rights* is now at the center of ethical discourse, and legal issues are obfuscating ethical issues. Within the area of scientific research there is an increasing awareness of the ethical implications of new developments on society. Issues such as pollution, overpopulation, the environment, nuclear power, computer technology, and bioengineering must be addressed within the context of ethics if our planet is to survive.

Ethics and Law

United States Supreme Court Justice Potter Stewart, in regard to the First Amendment, once said, "Just because we have the right to do something, doesn't make it the right thing to do." The law presupposes *minimal* ethical commitments, *not maximum* commitments. There is no question that law and ethics are intertwined, insofar as laws do encompass ethical principles and insofar as a *prima facie* ethical obligation exists to obey the law. On the other hand, this connection does not imply that they perfectly coincide, for the ethical principles encompassed may be objectionable, and *prima facie* obligations may, therefore, need to be overridden. Laws do not necessarily take into account all the intricacies that characterize concrete ethical judgements.

Ethical Reasoning–Reflective Deliberation

The essential component in reaching effective ethical decisions is practice. This means practice with ethically problematic situations via case studies, *not* mastery of a given set of principles and precepts by means of formal logical argument. It is a mistake to assume that ethical problems can always be effectively approached by plugging them into a theoretical framework and then deducing a solution. Theoretical principles and concepts may be used as tools that are often quite valuable for *identifying* issues and *suggesting* solutions, but ultimately it is a *process* of reasoning and deliberation on a case- by-case, situation-by-situation basis that brings about the resolution of ethical conflicts encountered during scientific research. The temptation to try to resolve ethical problems by means of a formula, or by applying a quantitative methodology, may be a natural inclination for the research scientist, but applying such methods to ethically problematic situations in which there are two or more conflicting duties or obligations does not solve such problems. Moral dilemmas are *cases of conscience*, and require us to examine all the possible underlying issues. Nonetheless, moral reasoning is not without basic standards or principles that must be addressed. Ethical deliberation is concerned with answering the following question: "What, all things considered, *ought* to be done?" (Benjamin, Curtis, p. 8). The nature of this question helps to distinguish ethical reasoning from other forms of critical thinking that take place in the practice of research.

Ethical reasoning must take into account such variables as the law, facts of the case, emotions, values, needs, and beliefs held by those individuals the decision will affect. Given this complexity, what "ought to be done" becomes very individualized and situational. This process of deliberation requires much more of us than adding new facts to existing information. It is much more than an intellectual exercise. It requires us to "feel" or empathize with an ethic such as justice through our emotions. The process requires us to explore things in new ways, while at the same time maintaining respect for the diverse moral views that exist in our pluralistic society. It also requires us to understand that moral issues *are not* simply matters of opinion. Issues regarding informed consent, justice, confidentiality, and equal access to social services cut across all of society's institutions and every profession.

Aristotle held that there is an important difference between ethical and scientific reasoning: Scientific reasoning requires bringing particulars under universal rules, whereas ethical reasoning, which he referred to as *practical* reasoning, requires evaluating often new and *unrepeatable* concrete situations and responding to them individually. Because ethically problematic situations are concrete and unique, ethical principles are helpful only as a foundation or starting point for deliberation. This is not to say that the process cannot have a structure or basic elements in the form of questions to assist us in arriving at a decision or conclusion. However, it does mean we must remember that there is no *one* ethical theory accepted by everyone, nor one set of criteria to evaluate different ethical positions. We *must* assume the premise that ethical dilemmas exist; they pervade life and cannot be ignored. Aristotle, in *Nicomachean Ethics*, addresses judging an ethical theory in the following way: Each person judges well what he knows, and is a good judge about that; hence the good judge in a particular area is the person educated in that area, and the unconditionally good judge is the person educated in every area. An in-depth presentation of the foundations of ethical theory and moral philosophy may be found through the Recommended Reading section at the end of this chapter.

Reflective Deliberation Applied to Case Studies

The model for ethical reasoning in this text combines critical thinking with reflective deliberation. It requires the cognitive skills of analysis, synthesis, comparison, and logic, in addition to a knowledge of facts, values, principles, and law. The model also incorporates personal wisdom, understanding, and experience. The application of the model is the case study. The case-study approach provides practice in applying ethics to real situations. This approach is referred to as *casuistry*, or particular case analysis, and it involves reflective deliberation of cases by addressing a set of questions. This approach requires us to adopt a broader, more flexible view of the situation than is required by theoretical analysis through formal logical argument. It demands an understanding of human needs and "the application of moral rules and other ethical considerations to new and more complex sets of circumstances, in ways that respect these human needs" (Jonsen, Toulmin, p. 342).

A Framework for the Reflective Deliberation of Case Studies

The following questions may be raised as a way to analyze cases:

1. What are the facts of the case?
2. Who are the stakeholders?
3. In what respect is the law relevant to the case?
4. What underlying imperatives, values, principles, standards, or codes are invoked?

5. To what extent are there competing or conflicting claims or duties?
6. What social and intellectual traditions are invoked for support?
7. What factors in the case are given weight, and which are excluded?
8. In what respect do other cases (examples) or precedents fit the circumstances of this case?
9. To what extent are the circumstances of this case exceptional or atypical?
10. What are the alternative positions (options) in this case?
11. What priorities do the options reveal? (i.e. what underlying values and commitments determine the priorities?)
12. What is or would be sacrificed for the sake of what is assumed to be the greater good in the various options?
13. What position do you support?

Justify your position by applying the following questions:

1. What are the underlying ethics of the position you support?
2. With which stakeholder(s) do you identify? Why?
3. What additional information could possibly change your position?
4. What *action* do you propose in support of your position?
5. What are the possible alternative courses of action available?
6. Are you comfortable with your position, *all things considered*?

Case Study 1

Ethics or Exploitation?
A Dilemma

A gene that stimulates ovulation in horses—but shuts off in winter—could be "turned back on," allowing horse breeding year round, according to a University of Florida scientist. "Horses usually begin breeding in the spring and give birth eleven months later, but most horse breeders would prefer foals born earlier in the year," said animal physiologist Dan Sharp at the University of Florida's Institute of Food and Agricultural Sciences Horse Research Center. "Our goal is to turn on the gene at the right time or prevent it from turning off, because once we do, breeders will be able to mate horses earlier," Sharp said.

Currently, only Mother Nature controls the gene, which is activated as mares are exposed to more daylight in spring, their natural breeding season. Thoroughbred breeder Bryan Howlett of Tartan Farms in Ocala said race horses born earlier in spring have a leg up on their competition and fetch a higher price on the open market. "Two-year-olds routinely have raced against horses eighteen months old or younger since 1753," he said, "when the Jockey Club of England declared January first the official birth date of all race horses born during the year. Thoroughbred horses born in January, February, or even March probably have 10 to 15 percent more value. Trainers think the extra development and time with the horses pays off," Howlett said.

Scientists from UF/IFAS and Case Western Reserve University in Cleveland, Ohio, now are studying how extra daylight prompts the mare's gene to produce luteinizing hormone. That hormone is responsible for ovulation, and mares normally secrete it in spring and summer. Scientists also have uncovered another hormone, called gonadotropin releasing hormone, which can signal the reproducing gene to "turn on or off." So far, researchers have successfully only "turned off" the LH-producing gene.

Source: Associated Press. *St. Petersburg Times* [Florida], 3 July 1994.

Analyze this case by applying the framework for reflective deliberation.

Questions for Discussion

1. What is so important to you that it could never be sacrificed for another end?
2. To what extent is scientific research objective and value-free?
3. In what respect are researchers accountable to society?

Recommended Reading

M. Benjamin, J. Curtis, *Ethics in Nursing*, 2nd. ed. Oxford University Press, New York 1986.

B. Gert, *The Moral Rules: A New Rational Foundation for Morality*. Harper, New York 1973.

A.R. Jonsen, S. Toulmin, *The Abuse of Casuistry: A History of Moral Reasoning*. University of California Press, Berkeley 1988.

A. MacIntyre, *After Virtue: A Study In Moral Theory*. University of Notre Dame Press, Notre Dame 1981.

P.W. Taylor, *Principles of Ethics: An Introduction*. Wadsworth, Belmont 1975.

Works Cited

M. Benjamin, J. Curtis, *Ethics in Nursing*, 2nd. ed. Oxford University Press, New York 1986.

A.R. Jonsen, S. Toulmin, *The Abuse of Casuistry: A History of Moral Reasoning*. University of California Press, Berkeley 1988.

J.R. Rest, "Can Ethics Be Taught in Professional Schools? The Psychological Research." In *Easier Said Than Done* (Winter 1988), pp. 22–26.

CHAPTER TWO

The Ethical Aspects of Scientific Research and Society

The right to search for truth implies also a duty; one must not conceal any part of what one has recognized to be true.

Albert Einstein

The Researcher in Society

ALMOST DAILY in this country the print media report a medical, scientific, or technological development that has created a situation warranting ethical consideration. The ethical implications of new scientific developments, such as the genetic links to certain diseases and the emergence of issues of priority and equity regarding organ transplants, genetic screening and engineering, foregoing life-sustaining treatment, confidentiality, conflicts of interest, and physician-assisted suicide have created dilemmas that cut across all levels of society. In addition, research scientists confront a sub-set of issues affecting the integrity of the research process itself. Such issues as accountability, amounts and patterns of funding, commercialization of research results, and the changing nature of collaborative efforts all have profound implications for researchers.

The system of internal checks and balances, which evolved in an environment far removed from the day-to-day activities of life outside the laboratory, is no longer protected by the self-correcting mechanisms that were once intrinsic to the scientific community. Now those involved in scientific research must assume an active role and accept responsibility for the consequences associated with their profession. Today the integrity of the research process goes beyond the hallowed walls of the university. However, by virtue of education and experience, scientists are better equipped than most people to understand, foresee, and assess the ethical implications of their work and assume responsibility for directing it toward humane ends.

Research has become a public enterprise. The recent allegations by Congress of research fraud in the tobacco industry are an example of the dilemmas scientists face when big business manipulates and controls the use of information derived from scientific research (see Case Study 2 C). Congress and public agencies are becoming increasingly vocal about what they perceive to be violations of public trust. The research process itself is under close scrutiny, and the outcome may well be more regulation. This, in turn, will affect the future direction of all research. The following chronology of ethics in research indicates that government regulation has been a consequence of misconduct by the researchers themselves. It is the responsibility of each individual scientist to recognize and uphold the standards of ethical science to ensure the integrity of the research process.

Chronology of Ethics in Scientific Research

1946   Nuremberg Code: A set of ten principles established as a result of revelations which emerged during the Nuremberg war crimes trials. First articulation of the concept of *informed consent.*

1950s   Small group of lawyers led by William Curran, Professor Emeritus of Legal Medicine at Harvard, urges scientists to seek ways to refine scientific practices used in human studies and to develop innovative principles to guide research.

1953   **James D. Watson and Francis H. C. Crick describe the structure of DNA.**

1960s   Civil rights movement increases social sensitivity to the needs of vulnerable groups.

1962   An amendment under the Food, Drug and Cosmetic Act requires investigators of *experimental drugs* to obtain a *formal consent* from subjects in their trials.

1963   Three physicians experimentally inject live cancer cells into twenty-two elderly and debilitated inpatients to determine whether such patients would reject the cells. Several physicians opposing this experiment express concern that the subjects never consented to participate. Case is brought before Board of Regents (BOR) of the State University of New York. Defending investigators contend that a doctor could *rightfully* withhold data that patients might find *threatening.* This position is not upheld by BOR.

1965 Pressure for reform increases: Henry Beecher, Distinguished Harvard Pharmacologist, publishes an article in the *New England Journal of Medicine* summarizing the *methods* of more than twenty-two studies, raising serious ethical questions about the *consent process* in human experiments.

1966 The National Advisory Council to the NIH develops guidelines requiring prior review of all research protocols on human subjects before a study can *begin.* Scientists now are required to obtain an informed consent from all research subjects and provide an *assessment of the possible harms and benefits* of the studies they propose. Peer review boards are established; these panels become known as *Institutional Review Boards,* or *IRBs.*

1967– Emergence of genetic research generates more public outcry concerning a *new* **1970** *eugenics movement.*

1973 Asilomar, California: One hundred scientists attend conference chaired by chemist Paul Berg to discuss the hazards of laboratory viruses.

Ability to splice and recombine different DNAs made known at a Gordon research conference in New Hampshire.

National Academy of Sciences convenes a committee to study *consequences* of making biological alterations; for example, joining DNA from animal viruses with DNA from bacteria.

1974 Second Asilomar conference held in September, chaired by Paul Berg. Guidelines are set up for recombinant DNA research. Biological and physical safeguards are to be used to contain new organisms generated in laboratories. Experiments involving highly pathogenic organisms would be deferred until more knowledge is gained. A statement is issued about the *ethical duties* of principal investigators; for example, they must inform staff members about the hazards of experiments before initiating them, make sure staff members are properly trained in containment procedures, and have procedures in place to monitor the health of staff members.

Congress passes National Research Act, establishes National Commission for the Protection of Human Subjects of Biomedical and Behavioral Research, and establishes the ethical principles for scientific research.

1975 Animal rights movement begins, spurred on by publication of Animal *Liberation* by Peter Singer.

1980s Animal rights groups appear, such as the Animal Legal Defense Fund, the International Society for Animal Rights, and the Coalition to End Animal Suffering in Experiments. By 1986, eighty bills dealing with the use of animals in research had been introduced in state legislatures throughout the U.S. *IRBs now required to examine all experiments involving animals from the viewpoint of humane treatment.*

1980s Big business becomes interested in the biological sciences. Massachusetts General Hospital receives two major industrial grants—one for $5 million from a German drug and chemical company for molecular biology research and a second for $885 million from a Japanese cosmetics company for study of the skin.

Biogen Company founded by Nobel Laureate Walter Gilbert.

1981 John Darsee, a medical scientist working in clinical and experimental cardiology, admits to falsifying data in one of his papers. Emory and Harvard Universities and NIH reveal his *fabrication* is more wide-spread. Concern that his co-workers had not detected flaws gives rise to the issue of *responsibilities of authorship* and casts doubt on the integrity of scientific literature.

By **1984**, industrial funds accounted for almost a quarter of the external support of university research in biotechnology, which included projects on genetically altered bacteria, cell and tissue cultures, DNA technology, monoclonal antibodies, and fermentation. *This growing relationship between scientists and industry gives rise to many ethical issues.*

1984 Robert Sinsheimer and colleagues at the University of California at San Diego attempt to create a genome sequencing institute. They fail to secure funding.

Charles DeLise and David Smith at the U.S. Department of Energy obtain funding for human genome project from the Department of Energy and NIH. Initial appropriation of $17 million from NIH.

1986 Nobel Laureate David Baltimore, as senior advisor to a project concerning genetic influences on the immune system, is challenged by Dr. Margot O'Toole, a member of the project, about the validity of the data. She loses her job. The case is brought before Congress and the NIH for investigation, and results in further examination of the responsibilities of authorship.

NIH and professional societies publish *corrective guidelines* delineating the responsibilities of authorship.

1989   NIH establishes the Office of Scientific Integrity (OSI). Its task is to delineate for investigators the responsibilities of handling and reporting possible *scientific misconduct* and investigating it. The OSI defines misconduct as *fabrication, falsification, or plagiarism.*

NIH mandates that all institutions seeking National Research Service Award training grants (NRSA) from the NIH and Alcohol, Drug Abuse, and Mental Health Administration are required to have a *program in the principles of scientific integrity* as an integral part of their efforts in researcher training.

National Center for Human Genome Research established with James Watson as director. Congress allocates $3 billion for the project, with $350 million going for the social, ethical, and legal implications of this research.

Federal guidelines instituted requiring that *ethics education* become a basic part of graduate studies.

1990   NIH issues policy and procedural guidelines governing the work of academic scientists

1992   NIH updates National Research Service Award requirements for instruction in the responsible conduct of research. All NRSA applications submitted after January 10, 1993, that do not contain a plan for such instruction will be considered incomplete, and will be returned to the applicant without review.

1993   NIH Revitalization Act requires the Secretary of Health and Human Services to develop a regulation to protect *whistle-blowers.*

NIH Revitalization Act replaces the term "scientific fraud" with "research misconduct" and requires that entities applying for PHS funds for any project or program that involves the conduct of biomedical or behavioral research provide assurance that they have in place a process to review reports of research misconduct.

1994   Councils of the National Academy of Sciences, the Institute of Medicine, and the Executive Committee of the Council of the National Academy of Engineering (NAS/IOM/NAE) issue a joint statement on research integrity.

Values in Science and Society

There is a relationship between religious, cultural, political, and economic values and scientific judgement. In other words, the attitudes and stereotypes we maintain in a societal context can influence choices and decisions made in the pursuit of scientific investigation. Therefore, it is most important for scientists to have self-awareness as well as an understanding of their religious values and the economic, political, and cultural influences that affect their choices and the resultant biases or distortions (Committee on Conduct, p. 8).

Scientific research works—and must work—against a background of assumed values. These values are derived from the three basic imperatives discussed in Chapter One: Human welfare, human justice, and human dignity. The values of life and happiness—maximizing the good, avoiding harm, safety, protection, and alleviating pain and suffering—are the values generated from the welfare imperative. The values of equality; fairness; trust and trustworthiness; acknowledgment (giving credit where it is due); honesty; truth and truthfulness; respect for laws, standards, customs, and policies; and accountability are the values of the justice imperative. From the dignity imperative come the values of freedom of choice, liberty, personal responsibility, and the right to privacy. These imperatives and their associated values form the foundation of all professional standards and codes of conduct. They are also the foundation of the responsible conduct of research.

Professional Standards and Codes of Ethics

Codes of ethics are formally-stated principles that embrace the values of a professional association and codify ethical conduct. They serve as guides, and exemplify the ideals of professional behavior. They also are used as a reference and justification for situations calling for ethical decisions. Included in the formal structure of such codes are statements regarding obligations to the profession; responsibilities to society; professional competence; relationships toward clients, colleagues, co-workers, and subordinates; confidentiality; and conflicts of interest. Codes of ethics generally do not cover such topics as public image or standards of compensation. As members of professional associations, individuals agree to uphold the standards set forth as a condition of membership. However, scientists working in research institutions are not bound by a common code of ethics. The ethics of scientific research have evolved from the traditional practice of the scientific method. The values of truth, honesty, integrity, objectivity, and fairness are practiced under an informal structure rather than under a written code. These traditions have been transmitted through example and instruction by mentors and supervisors as part of students' traditional training. The absence of a cohesive ethics code has created conflicts of obligation for scientists inside and outside the lab. Their right of autonomy may conflict with obligations to the public. Researchers can no longer "experiment" without considering the implications of their work outside the lab. The increasing importance of scientific discovery on human life necessitates constant diligence and careful scrutiny of these implications. A code of ethics for scientists would protect both the scientist and society.

The Nuremberg Code

The Nuremberg Code is the first document in contemporary society to address the ethics of experimentation and research involving human subjects. This code contains a set of ten principles established as a result of revelations which emerged during the Nuremberg war crimes trials. It contains the first articulation of the concept of *informed consent*. All ten principles embrace the three basic imperatives: Human welfare—beneficence, human justice—fairness, and human dignity—autonomy. In addition to formalizing the concept of informed consent, this document formalized the ethic of utilitarianism; that is, of doing the greatest good possible, taking into account not only the welfare of the individual but also all affected by the experiment or research. It also introduced a research procedure known as *risk/benefit analysis*, which has become a part of the informed consent protocol. One of the principles of the Nuremberg Code (Principle 6) states that "the degree of risk to be taken should never exceed that determined by the humanitarian importance of the problem to be solved by the experiment". This means that the benefit expected must *exceed* the expected harms (risks). In recent years, with the public's increasing awareness of research practices, Principle 6 has come under scrutiny and has created an ethical dilemma. The dilemma results from the inherent conflicting claims between the imperatives of autonomy and welfare/beneficence, or the conflict between the welfare of the individual and the interests (greater good) of society. The concept of *common good* is associated with *community*—the idea of a community joined in a shared pursuit of common values and goals; that is, a community of individuals whose own welfare is inextricably connected to the welfare of the whole. The common good therefore refers to those values which embrace well-being or prosperity for all. The concept of *public interest* is associated with *individual interest*—the idea of the aggregation of private interests of individuals who join together in an association dedicated to the pursuit of mutual advantage. The dilemma created for scientific researchers is this: How does one go about developing generalizable knowledge for the welfare of society (the greater good), while maintaining respect, privacy, and confidentiality of individual subjects, whose autonomy and protection from harm must also be maintained? The responsibility for the resolution of this conflict rests with the researcher; voluntary participation and informed consent remain ever present.

The Nuremberg Code

The proof as to war crimes and crimes against humanity

Judged by any standard of proof, the record clearly shows the commission of war crimes and crimes against humanity substantially as alleged in counts two and three of the indictment. Beginning with the outbreak of World War II, criminal medical experiments on non-German nationals, both prisoners of war and civilians, including Jews and "asocial" persons, were carried out on a large scale in Germany and the occupied countries. These experiments were not the isolated and casual acts of individual doctors and scientists working solely on their own responsibility, but were the product of coordinated policy-making and planning at high governmental, military, and Nazi Party levels, conducted as an integral part of the total war effort. They were ordered, sanctioned, permitted, or approved by persons in positions of authority who, under all principles of law, were under the duty to know about these and to take steps to terminate or prevent them.

Permissible medical experiments

The great weight of evidence before us is to the effect that certain types of medical experiments on human beings, when kept within reasonably well-defined bounds, conform to the ethics of the medical profession generally. The protagonists of the practice of human experimentation justify their views on the basis that such experiments yield results for the good of society that are unprocurable by other methods or means of study. All agree, however, that certain basic principles must be observed in order to satisfy moral, ethical, and legal concepts:

1. The voluntary consent of the human subject is absolutely essential. This means that the person involved should have legal capacity to give consent; should be so situated as to be able to exercise free power of choice, without the intervention of any element of force, fraud, deceit, duress, over-reaching, or other ulterior form of constraint or coercion; and should have sufficient knowledge and comprehension of the elements of the subject matter involved as to enable him to make an understanding and enlightened decision. This latter element requires that before the acceptance of an affirmative decision by the experimental subject, there should be made known to him the nature, duration,

and purpose of the experiment; the method and means by which it is to be conducted; all inconveniences and hazards reasonably to be expected; and the effects on his health or person which may possibly come from his participation in the experiment.

The duty and responsibility for ascertaining the quality of the consent rests on each individual who initiates, directs, or engages in the experiment. It is a personal duty and responsibility which may not be delegated to another with impunity.

2. The experiment should be such as to yield fruitful results for the good of society, unprocurable by other methods or means of study, and not random and unnecessary in nature.

3. The experiment should be so designed and based on the results of animal experimentation and a knowledge of the natural history of the disease or other problem under that study the anticipated results will justify the performance of the experiment.

4. The experiment should be so conducted as to avoid all unnecessary physical and mental suffering and injury.

5. No experiment should be conducted where there is an *a priori* reason to believe that death and disabling injury will occur; except, perhaps, in those experiments where the experimental physicians also serve as subjects.

6. The degree of risk to be taken should never exceed that determined by the humanitarian importance of the problem to be solved by the experiment.

7. Proper preparations should be made and adequate facilities provided to protect the experimental subject against even remote possibilities of injury, disability, or death.

8. The experiment should be conducted only by scientifically qualified persons. The highest degree of skill and care should be required through all stages of the experiment of those who conduct or engage in the experiment.

9. During the course of the experiment the human subject should be at liberty to bring the experiment to an end if he has reached the physical or mental state where continuation of the experiment seems to him to be impossible.

10. During the course of the experiment the scientist in charge must be prepared to terminate the experiment at any stage, if he has probable cause to believe, in the exercise of the good faith, superior skill, and careful judgement required of him, that a continuation of the experiment is likely to result in injury, disability, or death to the experimental subject.

Source: *Trials of War Criminals before the Nuremberg Military Tribunals Under Control Council.* Law No. 10, vol. 2. Washington, D.C.: Government Printing Office, 1949.

Risk/Benefit Analysis

As a way of protecting research subjects from harm, researchers have come to rely on an elaborate form of reasoning referred to as *risk/benefit analysis,* in which the projected benefits of the study multiplied by the probability of their occurring must *significantly exceed* the possible harms multiplied by their probability. This procedure is a statistical analysis that concludes with a *risk/benefit ratio,* which, when calculated, becomes part of the research protocol for informed consent. In order to assess the possible risks involved, the researcher must consider such things as social, economic, and legal risks, breach of privacy or confidentiality, potential physical or psychological harm, use of deception or coercion, and dissemination of invalid findings.

Informed Consent

Informed consent is a legal term referring to a person who, in possession of suitable information, grants authority to someone else to take actions affecting that person; in medical ethics and in scientific research involving human subjects, it indicates the subject's approval of a procedure or treatment based on acquisition and understanding of all relevant information, and it is grounded in the imperative of *autonomy.* It requires that the information be presented (that is, *communicated*) and *consent* obtained *prior* to proceeding with the treatment or research. Although the Nuremberg Code provided a framework for informed consent, it was not until 1966 that the U.S. Public Health Service (PHS) mandated that institutions receiving PHS research and training grants review all research involving human subjects to ensure that subjects' rights and welfare were protected, and that appropriate procedures were used to obtain their consent.

However, obtaining consent from research subjects is *not* just a matter of having them *sign* consent forms. The investigator must assume responsibility for having subjects *understand* the nature of the investigation, including the potential risks involved, as well as the purpose of the research and its expected outcome. Subjects must also understand that their participation is entirely *voluntary,* and that they have the right to *refuse* or *withdraw* from the study at any time. Many of the difficulties in the process of obtaining consent from subjects arise from the fact that those in question do not have a clear understanding of research and statistical methodology. This situation is exacerbated when the subject is either mentally impaired or has a life-threatening disease such as AIDS. A subject's eagerness to participate in finding relief from distress or even a cure may obscure his or her understanding of the risks involved and possible side-effects. In addition, the person may not understand the differences between scientific research and medical treatment, because the researcher may also be the attending physician. Procedures to ensure that research subjects have a clear understanding of the differences between research and treatment must become a part of every research protocol, and the essential component of this process is the way the information is *communicated* to subjects. Before proceeding with any research involving human subjects, whether it be scientific or non-scientific, the researcher should consider the following questions:

1. How will the subjects be chosen? (fairness – equity)
2. What are the risks and benefits? (welfare – beneficence)
3. How will the subjects be contacted? (autonomy – privacy)
4. How will the risks and benefits of the study be explained? (welfare – beneficence)
5. How will the protocol of informed consent, which includes voluntary participation and the right to withdraw, be explained? (welfare – autonomy – justice)

One of the most important aspects of the informed consent process is the *communication* that takes place between the researcher and the subjects. It is the responsibility of the researcher to be cognizant of the inherent ethics of the research *prior* to any communication with subjects in the study. This means that the researcher must be aware of and able to apply all of the following communication skills: appropriate eye contact, which shows respect; active listening, which allows participants to ask questions and express concerns; empathy, which is the ability to understand the subject's state of mind; tone of voice, which should be gentle, and should communicate respect and concern; and professional appearance, which signifies responsibility and commitment to the research. These skills will enhance and build a trusting relationship that allows the subjects to freely participate and exercise the right to autonomy, and they ensure that the principles of beneficence and justice will be maintained throughout the research process.

The documentation of informed consent should contain the following information:

- Informed Consent Form heading.
- Identification of the responsible institution(s) (e.g., The University of South Florida), principal and co-investigator(s), title of project, and funding agency/study sponsor.

- Human research and informed consent: Explanation of the general purpose of clinical research and the concept of informed consent.
- Purpose of research study: Explanation of the aims and purposes of this particular research, why the subject is being asked to participate, and the approximate number of subjects involved in the study.
- Description of the research study: Description of the procedures to be followed, and identification of the experimental procedures. Statement of the expected duration of the subject's participation. Explanation of use of a placebo agent or randomization-blinded study, if appropriate.
- Possible benefits of the research: Description of any benefits to the subject or to others that may reasonably be expected from the research.
- Risks: Description of any reasonably foreseeable risks, side effects, and/or discomforts to the subject (including likely results if an experimental treatment should prove ineffective). Statement, if applicable, that the particular treatment or procedure may involve risks to the subject or to an embryo or fetus if the subject is or becomes pregnant. Statement that there may be unforeseeable and unknown risks or side effects.
- Alternative procedures or treatment: Disclosure of appropriate alternative procedures or courses of treatment, if any, that might be advantageous to the subject, including the option not to participate and instead receive the usual standard-of-care therapy.
- Confidentiality: Statement describing how confidentiality of records identifying the subject will be maintained. Disclosure of who (FDA, subject's health-care providers, study sponsor) may have access to the records.
- Payment for participation: *If* the subject is to be offered payment as an inducement to participate, this information must be clearly stated on the consent form, including the payment policy under various contingencies (one-time fee or per visit, drop-out for medical reasons, etc.).
- Paying the costs of the research: Either a disclosure of any additional costs to the subject that may result from participation in the research study or a statement that there will be no additional costs to the subject for participation in the research study. Also a statement of any costs that will be paid by someone other than the subject or medical insurance.
- Injury resulting from the research: Statement as to whether or not compensation for injury is available from the sponsor and each participating institution.
- Voluntary nature of participation: Statement that participation is voluntary, and that refusal to participate or a subsequent decision to discontinue participation will not result in penalty or loss of benefits to

which the subject is otherwise entitled. Description of the consequences, if any, that would accompany a decision to withdraw from the research, and procedures for orderly termination of participation by the subject.

- New information arising during the research: Statement that any new information developed during the course of the research that may relate to the subject's willingness to continue participation will be provided to the subject, and that the subject's participation may be terminated by the investigator without regard to the subject's consent.
- Persons to contact: Whom to contact for answers to any questions the subject might have about the research or the research subject's rights. Statement should include the principal and/or co-investigators' daytime and evening telephone numbers.
- Consent: Statement to be signed by the subject indicating that the subject has been informed about the study, has had an opportunity to ask questions, has been given a copy of the consent form, understands the risks and alternatives, and freely chooses to participate. This statement must be provided in the subject's native language if the subject does not understand the language of the original document.
- Investigator statement: Statement to be signed by the investigator certifying that to the best of his or her knowledge the subject understands the nature, demands, risks, and benefits involved in participating in the study.
- IRB approval statement: Statement that the study and informed consent form have been approved by the IRB for a one-year period. Statement must be inserted at the bottom margin of the form, letter, or portion of the form that is to be retained by the subject.
- Summary of consent: Statement that summarizes the purpose, procedures, and voluntary nature of the study. Used only when the subject does not speak or understand the language of the original document; it must be translated into the subject's native language. Subject must be given a copy of this statement.
- Child's assent statement: Subject's agreement to participate in the study, if the subject is a minor.
- Paternal consent statement: Statement of paternal consent to research if the research involves pregnant women.
- Interpreter's statement: Statement of interpreter that the subject understands and consents to the research. To be added if the subject does not speak or understand the language of the document.

Institutions generally develop their own informed-consent statements and forms that may be used as models. Instructions to the investigator are usually in italics. However, such forms should always be adapted for the study in question, and should be written in simple, easily-understood language.

The Belmont Report

In 1971, as the public's growing awareness of abuses of patients' rights increased, the National Institutes of Health (NIH) set about establishing guidelines for protecting subjects in research. Subsequently, new federal regulations and a law establishing a National Commission for the Protection of Human Subjects emerged. The Commission issued what is now called the *Belmont Report*. This document set forth an ethical framework for assessing government research involving human subjects. The report identifies the ethical principles that should underlie the conduct of biomedical and behavioral research involving human subjects, and provides guidelines that assure that such research is conducted in accordance with those principles. The Belmont Report addresses such areas as boundaries between practice and research, application of ethical principles, disclosure, informed consent, selection of subjects, and assessment of risks and benefits (Belmont Report). Today, almost all institutions conducting such research apply these guidelines to their work, whether it is federally funded or not.

Confidentiality and Release of Information

Perhaps the most sensitive aspect of the research process from the perspective of the rights and welfare of subjects is the matter of confidentiality and release of information. This is particularly the case in AIDS research. Improper disclosure could have the most serious consequences for research participants by threatening family relationships, job security, employability, or the ability to obtain credit or insurance. In the light of these risks, special precautions must be taken to preserve confidentiality, and subjects must be informed about the limits of that confidentiality so they can make thoughtful, informed decisions as to whether to participate.

Each study must be designed with administrative, management, and technical safeguards to control the use and disclosure of information. Where identifiers are not required by design of the study, they are not to be recorded. If identifiers are recorded, they should be separated, if possible, from the data, and then stored securely, with linkage restored only when necessary for conduct of the research. No lists should be retained identifying those who elected not to participate. All participants should be given a clear explanation of how information about them will be handled, and no information may be disclosed without the subject's written consent. This is accomplished by use of a *release of information* form, a form that must clearly state who is entitled to see the records with identifiers, both within and outside the project. This statement must take account of the possibility of review of the records by the funding agency and by FDA officials if the research is subject to FDA regulations.

Conflict of Interest and Conflict of Commitment

Conflicts of interest and conflicts of commitment are defined as the predicaments arising when a person confronts two actions that cannot be ethically reconciled—competing loyalties and concerns with others; self dealing; outside compensation; or divided loyalties among, for example, public and/or professional duties and private or personal affairs. Recently, conflict of interest has come to refer to situations in which there are *financial interests* at stake, and conflict of commitment refers to situations in which competing loyalties and duties exist. The key issue for both conflicts, however, is the maintenance of *objectivity* and protection against *bias*. Hence, the regulations regarding these conflicts seek to minimize the probability that researchers will show bias in research when they or members of their families may personally profit from the results.

The emphasis on personal financial interest results from the fact that it is often more tangible and/or obvious than other forms of influence. This is not to say that other forms of influence are less damaging to the overall results of research endeavors. For example, sources of bias may infect the peer review process, or there may be a temptation to seek public recognition or exposure for personal advancement in one's career or other "perks" within the research community.

The ethical principles at stake even in the appearance of such conflicts are fairness, trustworthiness, and truthfulness. The principle of fairness relates to the issue of personal gain or gain at the expense of someone else's efforts. The responsibility for monitoring conflicts of interest rests with institutions receiving the grant or contract awards rather than with the federal agencies that give the awards and regulate the award conditions. The institution receiving the grant or contract is responsible for preventing, detecting, investigating, and correcting all conflicts of interest that involve individual researchers.

The NIH defines conflict of interest as "employees, consultants, or members of the governing bodies using their positions for purposes that are, or give the appearance of being, motivated by a desire for private financial gain for themselves or others such as those with whom they have family or business or other ties" (U.S. Department of Health). The NSF includes conflict of interest under the eclectic rubric of "misconduct in science" (National Science Foundation). In 1994 the NSF changed the term "misconduct in science" to "misconduct in research."

The Association of American Medical Colleges has a broader definition for conflict of interest:

> ... the term conflict of interest in science refers to situations in which financial or other personal considerations may compromise, or have the appearance of compromising, an investigator's professional judgement in conducting or reporting research . . . . The bias such conflicts may conceivably impart not only affects collection, analysis, and interpretation of data, but also the hiring of staff, procurement of materials, sharing of results, choice of protocol, and use of statistical methods . . . . Individual conflicts of interest in research arise in large part because of the interplay between a faculty member's personal and financial interests and the opportunity to conduct externally-funded research (AAMC Guidelines, p. 6).

Conflict of commitment occurs when obligations outside one's academic or administrative or employee mission take precedence over obligations to one's primary employer. These obligations may be in the same area of research, but in conflict with time commitments such as teaching, mentorship, and administrative responsibilities. In other words, whenever a commitment compromises one's internal work obligations, a conflict of commitment may have occurred. There may or may not be a financial interest involved. The difference between conflict of interest and conflict of commitment is that the former is more concerned with external interests that may compromise professional judgement and the reporting of scientific data, and the latter is more concerned with one's commitment to carry out professional responsibilities for an employer.

It is important to remember, however, that conflicts of commitment do not necessarily reflect negative or bad involvements, but rather indicate external obligations that conflict with internal obligations of employment. Potential conflicts of commitment may include such activities as consulting, holding office in professional societies, and community service. The point is that in these situations the activity itself may be good, but it may significantly interfere with responsibilities to the primary institution. Most institutions have disclosure policies that address potential conflicts of interest. Policies for conflicts of commitment are either nonexistent or vary widely among institutions and individual roles within the institution.

CASE STUDY 2 A

Parental Rights Versus Medical Judgement
A Dilemma

In what medical ethicists say could be an important test case of parents' rights to stop medical treatment for prematurely born infants, a Michigan doctor who disconnected his son's life-support system will stand trial on charges of manslaughter.

An hour after the baby's birth, the father, a dermatologist on the staff of the hospital, asked nurses to leave, then unhooked the life-support system, setting off an alarm. The baby was later pronounced dead, and the hospital then notified the police.

The physician's son was born fifteen weeks before the due date and weighed one pound, eleven ounces. The infant, considered at risk for brain damage, was given a 30–50 percent chance of survival. The case raises complex issues pitting parents' rights against medical judgement in a legal and ethical gray zone.

Although court rulings and state laws exist regulating whether parents have the right to withhold treatment from infants with Down's syndrome or spina bifida, the prosecutor and several medical ethicists said they knew of no precedent regarding severely premature infants.

State laws generally do allow parents to make medical decisions for their children, including the ending of life-support. The prosecutor in this case said, however, that those decisions were more commonly made after children had stayed on respirators for some time and doctors had had a chance to evaluate the prospects. The prosecutor claimed that the father appeared to make a unilateral decision to end life for his infant son. "Was his act in the best interests of the child? Had he allowed more medical tests, he and his wife would have been in a better position to evaluate the situation; but he took things into his own hands."

Many bioethicists say they prefer to start treatment and get information. Only then, if it appears that conditions warrant, would they stop treatment. However, they also say that in actual practice just the opposite happens, because it feels a lot different stopping treatment than not starting it.

Source: *New York Times* 3 Aug. 1994.

CASE STUDY 2 B

Politics, the Law, Informed Consent, and Confidentiality
A Dilemma

Ten thousand children born in New York with evidence of HIV infection were sent home with a dishonest bill of health. These children were tested for the disease at birth, but never received the preventive medicine available to adults, because of a law in New York that declares the HIV test to be confidential information because it reveals a mother's HIV status. Because the mothers do not consent to the tests for themselves or even choose to have them for their children, the test results cannot be revealed.

Last year Assemblywoman Nettie Mayersohn introduced a bill that would require the state of New York to disclose the results of its HIV screening to parents or guardians of newborns. At first, the chairman of the state health committee, Assemblyman Richard Gottfried, said he thought Mayersohn's bill "may very well be a practical way to save the lives of some newborns." But intense lobbying was initiated by the Gay Men's Health Crisis, which worries about erosions in the state's laws on confidentiality for HIV patients, and by the National Organization for Women (NOW). NOW has argued that testing the baby forcibly reveals the mother's HIV status and reduces the mother's control over her body and medical decisions. Gottfried has changed his mind and has blocked Mayersohn's bill by introducing one that calls for intense counseling before the baby is born, but continues to make testing voluntary for both mother and child. "Pre-serving a test based on consent is very important to keeping these infants involved in the health care system," says Gottfried. He also contends that knowing a baby's HIV status at birth hasn't helped the children prone to serious illnesses live longer.

Presently, the Centers for Disease Control reports that forty-four states blind test newborns for HIV virus, and none of these states have made any attempt to unblind the tests.

Source: Washington Post News Service. *Los Angeles Times* 18 Apr. 1994.

CASE STUDY 2 C

Conflicts of Interest: Whose Interest?
A Dilemma

Two former scientists working for Philip Morris U.S.A. told a U.S. House panel that their studies on rats more than a decade ago raised serious questions about the potential addictive nature of nicotine. They said the tobacco company suppressed their research and abruptly closed down their lab. "You cannot prove addiction from a rat, but you can say that further work is needed," Victor DeNoble told the House Energy and Commerce Subcommittee on Health and the Environment.

Steven Parrish, a senior vice president of Philip Morris, said that DeNoble had changed his opinions over time, making his findings more dramatic than they actually were. "That Dr. DeNoble has now conveniently changed his opinions does not change the facts of what his Philip Morris research showed," Parrish said.

From 1980 to 1984, DeNoble worked as a project leader for the behavioral pharmacology laboratory at the Philip Morris Research Center in Richmond, Virginia. Working with fellow scientist Paul Mele, DeNoble conducted studies on rats to find out whether they would self-administer nicotine intravenously if they could do so by pressing levers. They found out that the rats self-administered nicotine much more often than when a saline solution was made available in the same manner, and that nicotine has a positive reinforcing effect, DeNoble testified. And while it did not follow that the same patterns would be found in humans, it did raise questions that warranted further investigation, he argued.

The two scientists also told the House panel that they were led to believe they could publish work in journals while working for Philip Morris, but when they sought to publish a paper in a scientific journal, they were told by Philip Morris they could not. Other claims were: They were told to cancel a presentation of the findings they planned to deliver at a meeting of the American Psychological Association. They were summoned to brief top company officials on their findings, and one top Philip Morris executive allegedly asked, "Why should I risk a billion-dollar industry on rats pressing a lever?" They were told to continue their work despite the bans on discussing it, but then they were informed that their lab was being closed. When they returned to finish cleaning it, they found the lab emptied.

Source: Associated Press. "Tobacco Company Suppressed Their Work, Scientists Say." *St. Petersburg Times* [Florida] Health and Medicine Column 29 April 1994.

Questions for Discussion

1. To what extent do societal values influence a researcher's choices regarding scientific research endeavors?
2. Can you recall any situations in which economic or political priorities determined the direction of scientific research?
3. Why is it important to maintain the confidentiality of research subjects?
4. What are the essential differences between conflict of interest and conflict of commitment? How are they similar?

Recommended Reading

R.D. Alexander, *The Biology of Moral Systems*. Aldine de Gruyter, Hawthorne 1987.

Committee on the Conduct of Science, National Academy of Sciences, *On Being A Scientist*. National Academy Press, Washington D.C. 1989.

T. Kuhn, *The Structure of Scientific Revolutions*. University of Chicago Press, Chicago 1970.

Works Cited

The Association of American Medical Colleges, *Guidelines for Dealing with Faculty Conflicts of Commitment and Conflicts of Interest in Research*. Washington D.C. 1992.

The National Commission for the Protection of Human Subjects of Biomedical and Behavioral Research, *The Belmont Report*. U.S. Government Printing Office, Washington D.C. 1988.

Committee on the Conduct of Science, National Academy of Sciences, *On Being A Scientist*. National Academy Press, Washington D.C. 1989.

National Science Foundation, *National Science Foundation Grants Policy Manual*. 666, VI-7. Washington D.C. 1989.

U.S. Department of Health and Human Services, *National Institutes of Health Guide*. 18:32. Bethesda 1994.

CHAPTER THREE

Current Ethical Issues in Scientific Research and Society

> The great fault of all ethics hitherto has been that they believed themselves to have to deal only with the relations of man to man. In reality, however, the question is what is his attitude to the world and all life that comes within his reach. A man is ethical only when life, as such, is sacred to him, that of plants and animals as that of his fellow men, and when he devotes himself helpfully to all life that is in need of help . . . . The ethic of the relation of man to man is not something apart by itself: it is only a particular relation which results from the universal one.
>
> *Albert Schweitzer*
> *Out of My Life and Thought: An Autobiography*

Research and Social Responsibility

THE introduction of new knowledge, even when it is highly valued, frequently confronts public skepticism on the one hand and mythical optimism on the other. This is a common response to what may be perceived as a threat to the status quo—to that which is familiar in our day-to-day functioning, and to the way we have learned to think about life and living it. New knowledge may also threaten our values and systems of beliefs, particularly given the rapidity of scientific and technological development in society today. In the nineteenth century our lives were focused around small communities and those institutions that supported the values of individualism and the common good; the rights of the individual were one and the same with the common good.

As we in the United States approached the twentieth century, individual rights and the common good no longer had the same meaning. The frontier was closing, and industrialization was transforming our lives from a rural to an urban existence. This new way of life challenged the values and societal structures which had sustained us. The World Wars that followed, and the subsequent knowledge that humanity had a capacity to commit atrocities of unimaginable magnitude, once again forced us to examine our value systems and the meaning of existence.

Now, as we approach the twenty-first century, with the changes created through science and technology, we are faced with ethical dilemmas that are not easily reconciled. We are beginning to realize that future life on this earth depends on acceptance of the fact that we are all globally interdependent, and that there are universal values we must embrace if our planet is to survive. The new frontiers of research in science and technology can be the key to our survival if we keep in mind that the applications and uses of these new developments must be grounded in sound moral wisdom that must be explored and deliberated. The questions raised by this new knowledge trouble many people and leave them with unclear or contradictory and confusing ideas about what is right or ought to be done. Thus, it is becoming increasingly important for researchers to join in the ensuing moral dialogue regarding the ethical and social consequences of new knowledge.

The role of researchers in society is under scrutiny, and the integrity of the research process is suspect. Should scientists be held accountable, and to higher ethical standards? Have they betrayed the public trust? It would appear that this may be the case, as one examines current governmental regulatory efforts. For example, in *Washington Fax Weekly*, a weekly research and development digest, four out of seven recent reports contained issues which directly affect the research process: (1) Research Integrity Commission's meeting stirred by whistle-blower testimony, (2) Commission would provide an "environment that creates integrity" by setting detailed policy for research institutions, (3) President's research moratorium applies only to embryos fertilized for the sole purpose of research, and (4) Representative Dornan promises to lead charge against use of federal funds to support human embryo research (Washington Fax Weekly, p. 1).

Do research scientists have a responsibility to weigh the possible consequences of their research in terms of the expected *value* the results will hold for the greater societal good? Recently, public outcry and criticism were leveled against David Keen, a research chemist, who announced that he had developed two lethal bullets—one that could pierce bulletproof vests and another that could shred flesh. Keen defended his invention by stating that it would help citizens protect themselves against attackers and intruders. He is quoted as saying "the beauty behind it is that it makes an incredible wound, and that makes the target stop and worry about survival instead of robbing or murdering you. There's no way to stop the bleeding. I don't care who it hits. They're going down for good" (Associated Press). After Keen made this announcement he was forced to take his wife and twenty-month-old daughter to a secret place in another part of town because of anonymous death threats and hate calls. He said, "I'm being called names I can't believe; I'm a law-abiding citizen" (New York Times).

Some firearms experts questioned whether these bullets could do what Keen claimed, or whether they even exist. The National Rifle Association (NRA) called the story a hoax, and suggested he was a gun opponent trying to stir outrage. Keen denied that, although he acknowledged favoring laws to control firearms. Gun-control advocates called his comments "sick," and police feared the bullets would end up in criminals' guns, especially the version that could penetrate bulletproof vests. Keen responded to these protests by canceling production of the ammunition capable of penetrating bulletproof vests, but refused to cancel production on the "defensive rounds" flesh-ripping bullets. Both kinds are made of polymers, which are carbon-based plastics.

A New York senator and a congressman plan to introduce legislation banning any bullet capable of piercing bulletproof armor. Keen has applications pending before the Bureau of Alcohol, Tobacco and Firearms to manufacture and sell ammunition. His company, with more than thirty employees, began in 1985 to develop paints and coatings that help make aircraft invisible to radar. It had $3.6 million in sales in 1993, but, facing cutbacks in defense spending, Keen began working on ammunition as a way of diversifying. Last spring, the local Chamber of Commerce honored him as its top small-business executive of the year, citing in part the financial stability of his company, Signature Products. During the ceremony Keen was described as "just a normal businessman trying to run a business." His wife Jill, who is also a chemist, described him as "a little boy in a grown-up body; a hard worker who loves to roller-skate" (Associated Press).

This case raises several ethical questions regarding the researcher and social responsibility: 1. Does David Keen, as a professional chemist, have a duty or obligation to protect and advance the welfare or common good of society? 2. Does a conflict of interest exist that can affect his objectivity? 3. What and whose values are at stake in this case? 4. Does Keen have a right as a law-abiding citizen to run his company as he sees fit? 5. *Whose* rights should prevail in this case?

The Human Genome Project

The Human Genome Project (HGP) is another example of the need to understand societal and ethical consequences associated with recent developments in scientific research and technology. The HGP is considered one of the most significant scientific research projects ever undertaken. It is a fifteen-year project designed to characterize the human genome as well as genomes of selected model organisms. The goal is to determine the complete sequence of the three billion or so chemicals that constitute our genetic endowment, or genome—the DNA. That genome determines everything from the secrets of how humans develop and age to whether they will succumb to heart disease or have genetic predispositions for cancer and Alzheimer's disease. Critics of the project say that access to this knowledge might create an ethical mine field; they are concerned about how the resulting information will be used. Biologists, on the other hand, call it their equivalent to "landing on the moon." Because this project is recognized as one of the most important research ventures ever undertaken, having the most significant ethical implications for future generations, Congress has committed $3 billion to accomplish the task, with 5 percent of the amount dedicated to the identification of major issues related to the corresponding ethical, legal, and social implications (ELSI). What is perhaps even more amazing is that very few people are aware that a project of this magnitude exists, let alone understand the ethical ramifications.

Understanding our genetic destiny generates many ethical questions that involve complex decisions with significant social and economic consequences. Furthermore, we cannot assume that everyone will want to know about his or her genetic future, because the information generated from the HGP will force us to consider choices that will challenge the very essence of our being. The ethical issues connected with the HGP are many. They include disclosure, confidentiality and privacy, protection against discrimination, proprietary uses of genetic knowledge, abortion choices, and the issue of how much of this information is relevant to personal and public health and the greater good of society. All of these moral concerns have yet to be addressed fully.

We are at a time in scientific development in which the technology exists to genetically manipulate and "perfect" new life. The problem raised by this technology relates to the viewpoint that *facts* and *values* are one and the same; that is, choices that are made might be based on what is technologically possible (facts), rather than a consideration of consequences or what *ought* to be the greater societal good in the long run (values). And even when it is possible to differentiate between facts and values in consideration of the consequences, the problem of individual rights versus the common good must be reconciled. Presently, there is no agreement or mutual base of understanding on what this means and which is to prevail, because these concepts have many interpretations in a pluralistic society. Nor is there agreement on who has the authority to make such decisions, let alone to determine if the choices will in fact benefit our species.

The ELSI program of the Human Genome Project will not provide *answers* to the ethical dilemmas created by this undertaking; the solutions are in the domain of those who determine public policy. While it will certainly help articulate the ethical implications, its influence is limited. The HGP project has no real power to control the ways in which the results are applied (Hubbard, Ward, p. 59). Presently, twenty-seven ELSI studies are under way (National Center). The purpose of these investigations is to address the ethical issues raised as the HGP develops. This approach is new, and may well be the vehicle for generating necessary dialogue. The data gleaned from methodical investigations can provide sound foundations and input for discussion, particularly when this information is used for the establishment of public policy. However, before the information from the HGP is used in this way there will be moral issues in addition to the ones discussed earlier that must be deliberated, given that in a democratic society the principles of equality, fairness, and justice must be upheld. For example: As a society can we collect information which reveals individual differences to such a great extent and still uphold those principles, or will the information be used to discriminate against some who want access to opportunities or treatment? Will individuals acquire a distinct advantage or disadvantage based on their genetic predispositions? How then, will the "all people are created equal" doctrine be upheld? It is incumbent on all knowledgeable scientists to enter the dialogue and share the responsibility of carefully considering the consequences of this technology for the benefit of all.

Genetic Engineering and Technology

From the Manhattan Project (physicists, chemists) to the Apollo Program (engineers) and the Human Genome Project (biologists), scientists have maintained that powerful technologies can be used to improve the quality of life. Though this may be true, what is not being discussed is the potential impact these technologies will have on the way we think about ourselves. For example, the HGP will give researchers the necessary information to cure genetic disorders; but it will also enable them to change the human code through manipulation and intervention techniques. Through advanced recombinant DNA technologies scientists can now create new species, recreate old species, and replicate and multiply (clone) desired species. The ethical questions inherent in these endeavors are: Is there a line that should not be crossed even for scientific or other gain, and if so where is it? Who will be responsible for determining the boundaries of benefit and harm? Recently, a molecular biologist at a well-known research university refused a grant worth about $1.25 million awarded to him by the NIH. He had been conducting genetic research, and became convinced that such studies eventually could be harmful to the welfare of humanity. In a statement, he said that by returning the money he hoped to sound a warning about what he sees as the serious dangers of genetic engineering.

Scientists are already manipulating the genes of plants and laboratory animals. Some manipulated plants, such as a tomato that was genetically altered to control ripening, have been approved for general use. Certain human cell genes have been altered for treating specific disorders, but regulations formulated by NIH specifically forbid manipulations of genes that could affect inherited characteristics. Despite current regulation to control genetic research (which at this point is limited to research at universities), over the last decade hundreds and perhaps thousands of species have been genetically engineered. For example, pigs have been genetically medified to incorporate human growth genes to produce "super pigs" with more meat. Researchers are able to breed hens that no longer contain the brooding trait, which allows them to produce more eggs. And with the cloning of cattle, which is now commonplace, many biotechnologists feel that the creation of "perfect people" is the next logical step. An indication that this may well be the direction of the future is evident in the increase in sex-selection abortions. Presently, there are no regulations that limit the release of genetically engineered animals. However, a bill proposed by senator Mark Hatfield of Oregon would impose a moratorium on the patenting of certain human tissues and organs, human gene cells, and animal organisms. The bill, named the Life Patenting Moratorium Act of 1993 (S.387), lists concerns over conflicts of interest for researchers as among the several related issues:

> The rapid advances in biotechnology and biomedical research capabilities are creating a wide range of ethical, legal, economic, environmental, international and social issues, including concerns about the patenting of life forms, eugenics, genetic discrimination,

conflicts of interest for biomedical researchers, and genetic privacy considerations in insurance and employment (S.387, p. 2).

One of the most compelling fears behind the bill is the possibility of designing made-to-order human beings, and choosing the characteristics to be produced:

> Prominent members of the scientific community are discussing the possibility of the permanent alteration of the genetic code of human beings (referred to as germ-line research), yet Congress has not yet addressed the ethical, legal, economic, environmental, evolutionary, international and social implications of such experimentation (S.387, p. 2).

The origin of such concerns is traced back to April 7, 1987. On that day the U.S. Patent and Trademark Office approved a policy for the patenting of animals. Prior to that historic decision, no animal had ever been patented under U.S. patent laws. Since then, numerous such patents have been approved, and many more applications are pending. Issues surrounding the confidentiality of such genetic information, including that of purposefully developed lines and related patents, are among other concerns listed in the bill:

> (2) the Department of Commerce, the National Institutes of Health and the Department of State should work with the international community to develop international standards relating to the patenting of genetic information and access to such information (S.387, p. 6).

This bill would add a new prohibition on the patentability of certain biological inventions or processes to Chapter 10, Part II of Title 35 of the U.S. Code:

> (a) In General—No human being, human organ, organ subpart (genetically engineered or otherwise) or genetically engineered animal shall be considered patentable subject matter under this title. (b) Suspension—Except as otherwise provided in [this] section, during the 2-year period beginning on the date of enactment of this section. No
> (1) human tissue, fluid, cell, gene or gene sequence (genetically engineered or otherwise); or
> (2) animal or animal organism (genetically engineered or otherwise);
> shall be considered patentable subject matter under title. The prohibition may continue after such 2-year period pursuant to section 381(f) of the Public Health Service Act (S.387, pp. 3–4).

This bill would not affect any patents granted prior to its passage, and it clearly states that this is a temporary moratorium rather than a permanent prohibition:

> ... this section shall not be construed to detrimentally affect the rights of such individuals, but rather to maintain such rights until the expiration of the of the 2-year period (supra at p. 4).

The bill presently resides in the Subcommittee on Patents, Copyrights, and Trademarks. If action is taken, it may be blended into the next NIH reauthorization bill, which is where research ethics, informed consent, IRBs, conflict of interest, scientific misconduct, and related issues are often addressed for biomedical and behavioral research.

The fact that technology is evolving faster than the considerations that are necessary to weigh ethical consequences and societal costs is a basic dilemma. For example, in-vitro fertilization (IVF) is now seen as a normal procedure in reproductive science, and it is now technologically possible for a single infant to have as many as five contributing parents—sperm donor, egg donor, and surrogate mother, in addition to the mother and the father who rear the infant. To many ethicists, the only thing more frightening than unrestrained reproductive technology would be Congress' playing God and imposing limits. It appears that this bill is an attempt on the part of Congress to define the boundaries of genetic engineering. Research and social responsibility will no longer exist as separate self-contained entities.

More than twenty-five countries have commissions that set policy on reproductive technology. In Britain, cloning human cells requires a license the governing body refuses to grant. Violators face up to ten years in prison. In Japan, all research on human cloning is prohibited by guidelines that in that country's highly conformist society have the force of law. As genetic engineering advances, scientists exploring these frontiers will find that they must become ethicists as well.

Genetic Testing and Screening

The ethical dilemmas created by advances in genetic testing are complex and controversial. This situation will intensify as the HGP progresses and as information from the project reveals more and more about the human genetic code. And while these concerns presently focus on issues of privacy, confidentiality, proprietary use of information, availability of counseling, insurability, and protection against discrimination, it is important to remember that genetic testing will yield information that will define what it means to be human for generations to come. It is now possible to predict diseases for which there are currently no preventions or cures. For example, in the case of Huntington's disease (HD) it is possible to conduct both prenatal and presymptomatic tests for the condition, whose DNA marker was discovered in 1983, and although counseling is required for all individuals seeking diagnosis for HD (because it is part of an ELSI research project), recognition of the need for guidelines and regulations is just beginning to emerge.

In a report published by the Institute of Medicine (IOM), a panel of experts concluded that the fundamental ethical principles applied to genetic testing must be *voluntariness, informed consent, and confidentiality*. The panel also concluded that children should be tested *only* if a meaningful intervention is available, which means that under this standard HD tests could not be performed on children (Institute of Medicine). However, the American College of Medical Genetics (ACMG), which represents medical geneticists, disagrees with the IOM. They claim that not enough emphasis is being placed on the *values* and *benefits* that such technologies will engender for individuals and society. They further assert that the guidelines proposed by IOM do not take into account the necessity of research with vulnerable populations (for example, children) to develop and validate the effectiveness of new interventions. Members of the ACMG are fearful that regulations and policies will be determined by politicized bodies rather than research scientists and health professionals like themselves. In the IOM report the term *screening* is used to denote testing *large* populations—for example, all new babies at risk for a particular mutant gene that may lead to later disease. It uses the word *test* as in *genetic testing* to describe situations such as families, in which a *small* number of people is believed to be at high risk of carrying a harmful gene. Other key elements of the IOM proposal are:

(1) Screening of newborns should take place only when (a) there is clear indication of benefit to the newborn, (b) a system is in place to confirm this, and (c) treatment and follow-up are available for affected infants.
(2) Tests for fetal sex selection, through abortion, should be discouraged. More broadly, reproductive genetic services should not be used for eugenic goals, but to increase individual controls over reproductive options.
(3) The committee recommends caution in the use of and interpretation of presymptomatic or predictive tests.
(4) In the case of predictive tests for mental disorders the results must be handled with stringent attention to confidentiality to protect an already vulnerable population.
(5) Children should generally only be tested for genetic disorders for which there exists an effective curative or preventive treatment that must be instituted early in life to achieve maximal benefit (Institute of Medicine).

In the meantime, while genetic testing goes unregulated, many biotechnology companies around the world are developing a vast array of predictive genetic tests. These companies view the market for predictive disease testing as an extraordinary financial opportunity. However, there are many health professionals who question the ethical implications of genetic testing *without* the availability of counseling services. A research group in Cardiff, Wales, funded by the Commission of European Communities, has attempted to focus solely on the issue of genetic counseling and the impact of human genome analysis on clinical practice (University of Wales). In a study published in 1993, the six-member team found that there was a distinct global variation in the *values* surrounding such testing, and the different ways in which individuals evaluate the risks of disease or birth defects. The report notes that DNA technology has made possible in a twenty-five-year period an almost five-fold increase in the number of fully identified genetic disorders. Moreover, according to the report, the results of the worldwide Human Genome Project will enable not only the testing of individuals for the presence of a specific gene, but also carrier screening of large populations. Because every human being is estimated to carry between four and eight deleterious recessive genes, the potential of finding millions of people with increased risks for multifactorial genetic diseases (such as cancer and heart disease) is high. The report points out that "normality" and "health" may need to be redefined if such testing and screening is introduced on a widespread basis.

The report also points out that there exists a wide diversity among European nations in approaches to and the availability of genetic counseling, but concludes that such counseling is important to protect *autonomy, personal integrity, and privacy*. The report further recommends that the traditional presumption in favor of *confidentiality*, while allowing discretion to *disclose information* for compelling reasons (such as avoiding harm to spouses/partners and other relatives), be maintained by counselors. The report concludes by noting that people undergoing testing have wide variations in their perceptions of risk, and in the significance they attach to the stakes involved in such decisions, particularly with reference to the decision to have children who may be born with birth defects. The report also alludes to the fact that governments also differ in the extent to which they feel they should take steps to further efforts toward disease prevention.

This report succinctly points out the ethical concerns in genetic testing, as well as the complexity of the solutions. Whatever decisions are made in the future about the use of information obtained by genetic testing, those decisions will have considerable social and economic consequences. The ethical uses of this information ought to be an immediate and primary concern of all researchers and health professionals.

Embryo Research

In a White House press release on December 2, 1994, President Clinton said that he did not believe federal funds should be used to support the *creation* of human embryos for the *sole purpose of research*, and he directed the National Institutes of Health not to allocate any resources for such research. To ensure that advice on complex bioethical issues that affect our society can continue to be developed, the president also recommended that the administration establish a National Bioethics Advisory Commission (President on NIH).

In September 1994, a panel of nineteen experts from NIH recommended that *under strict guidelines*, the government should support embryo research (Hoke, p. 1). The types of research considered by the panel for funding involve fertilized human ova, either in vitro or flushed from the womb before implantation in the uterine wall. Examples of procedures reviewed include embryo biopsies for genetic testing, toxicological and nutritional studies, and various protocols aimed at a better understanding of the reproductive process. Scientists define the embryo stage as extending into the eighth week of gestation, at which point the embryo is termed a fetus. Only very-early-stage embryos were considered by the panel for possible use in research. In addressing the ethics of performing research involving embryos, the panel concluded that "the preimplantation human embryo warrants serious moral consideration as a developing form of human life, but does not have the same moral status as infants and children" (Hoke, p. 6). The panel advised that most investigations be discontinued about fourteen days after fertilization, when the so-called primitive streak appears.

This recommendation was followed by an immediate public outcry, with opponents claiming that the primitive streak may have no special *ethical* significance as a stopping point on the developmental continuum from conception to birth. This issue is still being debated. The panel, in defense of their position, claim that in forming their recommendations they sought to create defensible public policy, a process that included public testimony at panel sessions before the report's release, as well as solicitation of public comment. The content of letters received by the panel during this process prompted the panel to request NIH to undertake a concerted public education effort to explain the complex scientific and ethical issues involved in human embryo research. To this end, NIH held an extended special briefing for science writers a week prior to the report's release, and prepared comprehensive educational material for distribution with the report.

This NIH advisory panel concluded that there were three principles guiding their recommendations: One, that the promise of *benefit* from doing research on the human embryo was *significant* and carried great potential for couples, families, and individuals. Two, that the preimplantation embryo warrants serious *moral consideration*, but does *not* have the same moral status as infants and children. Three, that federal funding and regulation will help bring about consistent ethical and scientific review of proposals to conduct research on the human embryo. Though most scientists accept these principles, there is widespread criticism, primarily from non-scientists and members of conservative religious or political groups also opposed to abortion. The question of when an embryo becomes uniquely human and therefore no longer an appropriate research subject puts scientists in a double-bind situation—one must do the research to get the answer.

At the same time, the National Advisory Board on Ethics in Reproduction (NABER), a privately funded, nonprofit group of thirteen professionals in ethics, medicine, law, religion, and public policy, is working on ways to combine existing research and public discussion in an attempt to develop guidelines for research dealing with human reproduction. As outlined in their first report, the following ethical issues will be addressed:

- Control and access: Does every couple have a right to have a family, using whatever means available? If so, is infertility an illness? Will government help those who can't afford the high costs? If not, will government enforce laws against it? Who should have access to genetic screening and for what purpose?
- Coercion and commercialism: Will women's choices actually be lessened because of pressure to have babies, now that the new techniques are here? Will couples be forced to use prenatal diagnosis because of public censure if they have a disabled child? Will surrogate motherhood become not only a profession, but a new way of exploiting women? Will humans become commodities?
- Genetics and the perfect person: What genetic tests should be offered—or required? Who will decide what defects are serious enough to screen out? Shall industries use genetic information (like a predisposition to heart disease) to deny employment? Will widespread screening make us less tolerant of people with disabilities?
- Parents and helpers: Do test-tube babies and surrogate mothers strengthen the family or threaten the whole idea? Who should have priority in a dispute, a birth mother or the rearing mother? Will children suffer psychologically or physically because they are born of these techniques? Should donors be anonymous? Have male donors and female donors been treated with equal fairness?

- Health-care professionals: Many of the new techniques are not medically indicated—they don't cure an illness. Are physicians obligated to perform them? How should the powers and responsibilities of the patient and the professional be balanced?
- Manipulating embryos: Is the newly fertilized pre-embryo, microscopic in size, a human being? What limits should be set for creating, manipulating or experimenting with it? Amid rising health costs, should society demand that embryos carrying certain disorders be discarded? (National Advisory Board)

As the report says, some ask, "At what point does medical genetics become eugenics?" (the search for genetic perfection). In a democracy, public policy is not simply imposed on citizens by a higher authority. Proper public policy develops through a public process rooted in public participation. Public participation in turn involves a multitude of interests, therefore it must strive for a balance among divergent interests—a balance sufficient to obtain and justify public support. The researcher also has a responsibility to participate in the deliberation process of determining social policy.

CASE STUDY 3 A

Baby to Have Six Parents in Unusual Surrogate Plan

When John and Jo-Lynne Seeger's daughter is born in September, her future sister will be her mother. And her birth mother's father will be her adoptive father. Altogether she will have six parents.

Nearly two years ago the Seegers, who couldn't have a child of their own, adopted a baby boy, Nathaniel. The couple then set out to adopt another child and wanted the child to have the same genetic parents as Nathaniel. They reasoned the two children would then have similar traits and a common background. The birth parents, who are unmarried and live in different states, declined to conceive a second child but offered to donate eggs and sperm. But while the embryos could be conceived in a laboratory, Jo-Lynne was unable to carry them to term.

That's when John Seeger's adult daughter from a previous marriage, Jolene Stone, thirty, volunteered to be the baby's surrogate mother. Stone, a mother of five and stepmother of two, will give birth to her future adoptive sister in early September. Stone will be the baby's legal mother when she is born. Stone's husband, David, will be the baby's legal father. The Seegers will then adopt her, shifting the family's relationships. Stone will become the child's half-sister. Jo-Lynne, forty-two, and John, fifty-three, will be her parents.

Source: "Baby to Have Six Parents in Unusual Surrogate Plan." *St. Petersburg Times* [Florida] 21 Aug. 1994.

CASE STUDY 3 B

Should Children Be Told If Genes Predict Illness?

In their fevered race to isolate the breast cancer gene, researchers have often discovered through indirect tests which relatives of affected women also carried the gene—and thus an 85 percent chance of developing the disease. Many of those relatives were children.

The researchers faced a troubling question: Should they tell the parents and children what they knew? To the families' shock and dismay, some decided not to. Many parents argued, to no avail, that they had a right to medical information about their children. But many of the geneticists felt that families should not be told because nothing could be done to prevent the disease. Increasingly, geneticists are asking, "Whose right is it to make that decision?"

The question is becoming pressing as researchers find more ways to identify who is at risk of developing painful, deadly diseases. The problem, scientists and ethicists say, is that far more is known about predicting these ailments than preventing or treating them. And they worry that the knowledge of future illness could be too great a burden for some children and parents to bear.

Source: Kolata, Gina. "Should Children Be Told if Genes Predict Illness?" *The New York Times*, 26 Sept. 1994, 1.

Questions for Discussion

1. Can you recall any experience in your life in which your right to be autonomous conflicted with what may be considered the greater good of society?
2. Can you recall any situations where societal stereotyping influenced scientific experimentation?
3. Do research scientists have a responsibility to weigh the possible consequences of their research in terms of the expected value the results hold for the greater societal good?
4. Should we ever limit on economic grounds what is spent on medical care for those who are very ill or extremely old?

5. What part do you believe moral views should play in determining public policy and legislation?
6. Is there a line that should not be crossed for scientific or other gain, and, if so, where is it?

Recommended Reading

C.F. Cranor (ed.), *Are Genes Us? The Social Consequences of the New Genetics.* Rutgers University Press, New Brunswick 1994.

R.C. Deegan, *The Gene Wars.* Norton, New York 1994.

R. Hubbard, E. Wald. *Exploding The Gene Myth.* Beacon, Boston 1993.

D.J. Keveles, L. Hood (eds.), *The Code of Codes: Scientific and Social Issues in the Human Genome Project.* Harvard University Press, Cambridge 1994.

Works Cited

Associated Press. News Release 30 Dec. 1994.

F. Hoke, *New Funds Possible for Embryo Research.* The Scientist 28, Nov. 1994, pp. 1–7.

R. Hubbard, E. Wald. *Exploding The Gene Myth.* Beacon, Boston 1993.

Institute of Medicine. *Assessing Genetic Risks: Implications for Health and Social Policy.* National Academy Press, Washington D.C.: 1994.

National Advisory Board on Ethics in Reproduction, *Report: Issues We Face.* Washington D.C. Nov. 1994.

National Center for Human Genome Research, National Institutes of Health, Ethical, Legal and Social Implications Branch, *Active Research Grants as of May 31, 1994.* Bethesda.

New York Times, 1 Jan. 1994.

President on NIH and Human Embryo Research. White House Summaries, 2 Dec. 1994.

Times Wire Service, 18 Nov. 1994.

University of Wales, College of Cardiff, *Centre for Applied Ethics. Report.* 1993.

Washington Fax Weekly R&D Digest, 9 Dec. 1994, pp. 1–5.

CHAPTER FOUR

The Ethics of Data Management

Nullius in Verba—On No Man's Word

Motto of the Royal Society of London
(First scientific society in the world—founded in 1660)

Responsible Research Practices

ETHICAL DATA management is central to ensuring the integrity of the research process. Although the methodology of gathering and recording data may differ among the individual disciplines of science (social science research depends on interpretation and field work; laboratory science has variable potential for replication due to maturation effects, testing effects, social change, or expensive research designs), there are basic values which must be upheld to protect the credibility and reputation of any researcher. There is a fine line between "sloppy science" (questionable research practices) and *misconduct*. Over the past decade newspapers around the world have reported astounding incidents of misconduct in scientific research that have undermined the integrity of the research process, prompting government agencies and professional associations to respond by setting up guidelines for the practice of responsible research.

Objectivity

Though scientists must apply methodological value judgements (choice of statistical tests, choice of sample size, interpolation of missing data, and interpretation of data) in order to advance the research process, interpret results, and draw conclusions, there are other values intrinsic to the research process itself. All the values discussed in previous chapters are important, but the value of *objectivity* is a keystone ethic in scientific research. Objectivity in research means avoiding bias. It is the duty of researchers to promote the *unbiased* use of research. One method of promoting unbiased research is to use women and minorities (as well as white males) equally as research subjects. Objectivity also means *self-scrutinizing* and *questioning* to prevent deception or suppression of potential dangers. Objectivity means making decisions about which data to use or ignore. Objectivity is crucial to the choices one makes in the process of sharing, reporting, and publishing results. Objectivity necessitates *veracity*. To be objective one must understand the value of *truthfulness*, which is a canon of research. Lack of objectivity in research obscures its basic purpose.

Competence

Because scientific research covers a broad range of disciplines, and replication is uneven across disciplinary areas, no individual researcher can be expected to know all the research methodology of all the other disciplines. However, that does not absolve the researcher from the basic requirement of *competence*. Researchers have a professional obligation to use only those methodologies in which they have been trained or with which they have had prior experience. The delivery of timely and reliable data is possible only when researchers stay within the areas of their training and abilities. Also, for scientists to remain competent within their respective disciplines, they must make a lifelong commitment to understanding current developments and advances through continuing education and training.

Research Data

The term *research data* applies to many different forms of scientific information, including raw numbers and field notes, machine tapes and notebooks, edited and categorized observations, interpretations and analyses, derived reagents and vectors, and tables, charts, slides, and photographs. Research data are the basis for reporting discoveries and experimental results. Scientists traditionally describe all the methods used for an experiment, along with appropriate calibrations, instrument types, the number of repeated measurements, and particular conditions that may have led to the omission of certain data in the reported version of their work. The standard practice is to provide information that is sufficiently complete so that another scientist can repeat or extend every experiment.

When a researcher communicates a set of results and a related theory or interpretation in any form (at a meeting, in a journal article, or in a book), it is assumed that the research has been conducted as reported. It is a violation of the most fundamental aspect of the scientific research process to set forth measurements that have not in fact been performed (fabrication), or to ignore or change relevant data that contradict the reported findings (falsification) (Panel on Science 1, p. 47).

In response to public scrutiny and accusations of misconduct in research, institutions and universities are planning major educational initiatives and creating guidelines to foster research integrity. In 1990, the NIH issued its first such guidelines, *Guidelines for the Conduct of Research* (U.S. Department of Health, p. 6). These guidelines state general principles which NIH scientists are expected to follow in their practice of research, and provide a framework to assist both new and experienced investigators as they strive to safe-guard the integrity of the research process (U.S. Department of Health, p. 1). The following excerpts from the NIH guidelines exemplify the highest standards and principles of ethical research practices in data management:

- Notebooks: The results of research should be carefully recorded in a form that will allow continuous access for analysis and review.
- Attention should be given to *annotating and indexing* notebooks and *documenting* computerized information to facilitate detailed review of data.
- All data, even from observations and experiments not directly leading to publication, should be treated comparably.
- All research data should be available to scientific collaborators for immediate review, consistent with requirements of confidentiality.
- Investigators should be aware that research data are legal documents for purposes such as establishing patent rights or when the veracity of published results is challenged, and the data are subject to subpoena by congressional committees and the courts.
- Research data, including the primary experimental results, should be retained for a sufficient period to allow analysis and repetition by others of published material resulting from those data. In some fields, five or seven years are specified as the minimum period of retention, but this may vary under different circumstances.
- Research data and supporting materials, such as unique reagents, belong to the National Institutes of Health, and should be maintained and made available, in general, by the laboratory in which they were developed. Departing investigators may take copies of notebooks or other data for further work. Under special circumstances, such as when required for continuation of research, departing investigators may take primary data

or unique reagents with them if adequate arrangements for their safe-keeping and availability to others are documented by the appropriate institute, center or division official.

- Data management, including the decision to publish, is the responsibility of the principal investigator. After publication, the research data and any unique reagents that form the basis of that communication should be made available promptly and completely to all responsible scientists seeking further information. Exceptions may be necessary to maintain confidentiality of clinical data or if unique materials were obtained under agreements that preclude their dissemination (U.S. Department of Health, p. 6).

Ownership and Authorship

Misconceptions regarding ownership of data are common among researchers, though universities and research institutions have policies and guidelines for authorship. In the NIH Intramural Programs, research data are considered to be the property of the Institutes, not individual researchers (Panel on Science 1, p. 48). Many scientists, however, regard their own research as their personal property, rather than the property of the institution where they are employed. This misconception can lead to acts that may be interpreted as misconduct. Often, scientists' interpretations of research ownership come from their early training experiences, which may not have been models of responsible conduct, and may not represent appropriate examples of the special obligations and duties scientists have toward the institutions that employ them, nor toward the granting agencies that fund their research.

The obligation to maintain an environment of *openness* by the sharing of data, peer review, and co-authorship is an essential component in maintaining the integrity of the research process, and this goes beyond what may be legally required in day-to-day business practices outside the institution.

Publication Practices

The National Institutes of Health have developed guidelines for publication practices in the Intramural Research Program, and these may be used as a framework to understand the importance of publishing in the overall research process. Publication of results is an integral and essential component of research (U.S. Department of Health, p. 8). Other than presentation at scientific meetings, publication in a scientific journal should normally be the mechanism of choice for sharing new findings. Exceptions may be appropriate when serious public health or safety issues are involved. Although appropriately considered the end point of a particular research project, publication is also the beginning of a process in which the scientific community at large can substantiate, correct, and further develop any particular set of results (U.S. Department of Health, p. 7).

Timely publication of new and significant results is important for the progress of science, but fragmentary publication of the results of a scientific investigation, or multiple publications of the same or similar data, are inappropriate. Each publication should make a unique and substantial contribution to its field. As a corollary to this principle, tenure appointments and promotions should be based on the importance of scientific accomplishments and not on the number of publications in which those accomplishments have been reported (U.S. Department of Health, p. 8).

Each paper should contain sufficient information for the informed reader to assess its validity. The principal method of scientific verification, however, is not review of submitted or published papers, but the ability of others to replicate the results. Therefore, each paper should contain all the information that would be necessary for the author's scientific peers to repeat the experiments. Essential data that are not normally included in a published paper (for example, nucleic acid and protein sequences and crystallographic information) should be deposited in the appropriate data base. Common practice regarding replication also requires that any unique materials (monoclonal antibodies, bacterial strains, or mutant cell lines, etc.), analytical amounts of scarce reagents, and unpublished data (e.g., protein or nucleic acid sequences) that are essential for repetition of the published experiments be made available to other qualified scientists. It is not necessary to provide materials that may be limited in supply (U.S. Department of Health, p. 9).

Authorship

Authorship refers to the listing of names of participants in all communications—oral and written—of experimental results and their interpretation. Authorship is the fulfillment of the responsibility to communicate research results to the scientific community for external evaluation (U.S. Department of Health, p. 10).

Authorship is also the primary mechanism for determining the allocation of credit for scientific advances, and thus the primary basis for assessing a scientist's contribution to developing new knowledge. As such, it potentially conveys great benefit, as well as responsibility. For each individual the privilege of authorship should be based on a significant contribution to the conceptualization, design, execution, and/or interpretation of the research study in question, as well as a willingness to assume responsibility for that study. Individuals who do not meet these criteria but who have assisted in the research by their encouragement and advice, or by providing space, financial support, reagents, occasional analyses, or patent material, should be acknowledged in the text but not listed as authors.

Because of the variation in practice among disciplines, a universal set of standards regarding authorship is difficult to formulate. It is expected, however, that each research group and Laboratory or Branch will freely discuss and resolve questions of authorship before and during the course of a study. Further, each author should fully review material that is to be presented in public forums or submitted (originally or in revision) for publication. Each author should be willing to support the general conclusions of the study (U.S. Department of Health, p. 10).

The submitting author should be considered the primary author with the additional responsibilities of coordinating the completion and submission of the work, satisfying pertinent rules of submission, and coordinating responses of the group to inquiries or challenges. The submitting author should ensure that the contributions of all collaborators are appropriately recognized and must be able to certify that each author has reviewed and authorized the submission of the manuscript in its original and revised forms. The recent practice of some journals of requiring approval signatures from each author before publication is felt to be useful in fulfilling that responsibility (U.S. Department of Health, p. 11).

Peer Review and Privileged Information

Peer review can be defined as expert critique of a scientific treatise (such as an article prepared or submitted for publication), a research grant proposal, a clinical research protocol, or an investigator's research program, as in a site visit. Peer review is an essential component of the conduct of research. Decisions on the funding of research proposals and on the publication of experimental results must be based on thorough, fair, and objective evaluations by recognized experts. Therefore, although it is often difficult and time consuming, scientists have an obligation to participate in the peer-review process; in doing so, they make an important contribution to science (U.S. Department of Health, p. 12).

Peer review requires that the reviewer be expert in the subject under review. The reviewer, however, should avoid any real or perceived conflict of interest that might arise because of a direct competitive, collaborative, or other close relationship with one or more of the authors of the material under review. Normally the existence of such a conflict of interest would require a decision not to participate in the review process, and to return any related material unread. The review must be objective. It should be based solely on scientific evaluation of the material under review within the context of published information, and should not be influenced by scientific information not publicly available (U.S. Department of Health, p. 13).

All material under review is privileged information. It should not be used to the benefit of the reviewer unless it previously has been made public. It should also not be shared with anyone unless necessary to the review process, in which case the names of those with whom the information was shared should be made known to those managing the review process. Material under review should not be copied and retained or used in any manner by the reviewer unless specifically permitted by the journal or reviewing organization and the author (U.S. Department of Health, p. 13).

Rights and Responsibilities of Peer Review

It is only through effective peer review that scientists and scholars can guarantee the highest standards of their profession. The University of Michigan medical school offers the following as their guidelines for maintaining the quality and integrity of research:
- Rights: Anonymity should be guaranteed to the reviewer.
- Responsibilities: Accept material for review only if qualified to do so (competence).
- Preserve the integrity of the review process (objectivity).
- Maintain *confidentiality* at all times.
- Ensure *impartiality* by identifying any potential *conflicts of interest*.
- Document the basis for negative evaluations (objectivity, honesty, respect, truthfulness).
- Strive to be *reasonable* and *fair*, particularly in requesting additional data.
- Submit reviews in timely fashion (respect) (Panel on Science 2, p. 145).

Questionable Research Practices

The Panel on Scientific Responsibility and the Conduct of Research at the National Academy of Sciences identifies *questionable research practices* as actions that violate the traditional values of research, and actions that may be detrimental to the research process (Panel on Science 1, p. 28). Questionable research practices include activities such as the following:
- Failing to retain significant research data for a reasonable period;

- Maintaining inadequate research records, especially for results that are published or are relied on by others;
- Conferring or requesting authorship on the basis of a specialized service or contribution that is not significantly related to the research reported in the paper;
- Refusing to give peers reasonable access to unique research materials or data that support published papers;
- Using inappropriate statistical or other methods of measurement to enhance the significance of research findings;
- Inadequately supervising research subordinates, or exploiting them; and
- Misrepresenting speculations as fact, or releasing preliminary research results, especially in the public media, without providing sufficient data to allow peers to judge the validity of the results or to reproduce the experiments (Panel on Science 1, p. 28).

When the Royal Society of London was founded in 1660, it took as its motto "Nullius in Verba," which means *on no man's word*. In essence this means that the final authorities in science are the data. The practice of responsible research is contingent on the acknowledgment (either implicit or explicit) that data must be reliable; they must be gathered, analyzed, interpreted, replicated, and shared objectively and competently. Data must withstand the scrutiny of peer review, and must be appropriately disseminated. The practice of responsible data management is the key element underlying all research advances.

CASE STUDY 4

Questionable Research Practices

It is now widely accepted that scientific researchers who develop new DNA sequences (clones), viruses, antibodies, or other reagents using federal funds will share the reagents with other researchers in a timely manner at reasonable cost (usually free). However, it is common experience that researchers ignore the requests they receive so that they can themselves do all the important experiments with that reagent without competition from other researchers. According to one story, scientist A requested a novel form of recombined virus from scientist B, who had just published some exciting results using this novel recombined virus. After three letters requesting the virus, scientist B finally wrote back to scientist A and complained that, because of a freezer break-down, they no longer had a sufficient amount of the novel recombined virus to share with others, but that as soon as new stocks were grown, scientist B would ship some to scientist A (the scientific equivalent of "the check is in the mail"). As soon as scientist A received the letter, he cut it up into small pieces and cultured it in petri dishes. In at least ten of the dishes he was able to grow the virus he had requested. Evidently the virus was so abundant in the laboratory of scientist B that even the stationery became "infected" (probably from scientists B's hands).

Source: Morgan, David G. Letter. March 1995.

Note: This is reported to be a true story, although Dr. Morgan had no real measure of its veracity.

Questions for Discussion

1. To what extent do pressures to produce and publish influence the exercise of responsible research practices?
2. What are the benefits and disadvantages of institutional guidelines for the conduct of research?
3. To what extent does peer review ensure the integrity of the research process?
4. What is the difference between *questionable research practices* and misconduct?
5. How is authorship credit best determined?

Recommended Reading

P. Medawar, *Advice To A Young Scientist*. Harper, New York 1979.

W. Broad, N. Wade, *Betrayers of the Truth: Fraud and Deceit in the Halls of Science*. Simon, New York 1982.

A. Kohn, *False Prophets*. Blackwell Scientific, New York 1986.

D. Nelkin, *Science As Intellectual Property: Who Controls Scientific Research?* Macmillan, New York 1984.

H.M. Kanare, *Writing The Laboratory Notebook*. American Chemical Society, Washington D.C. 1985.

Works Cited

Panel on Science, Engineering, and Public Policy, *Responsible Science: Ensuring the Integrity of the Research Process*, 2 vols. National Academy Press, Washington D.C. 1992—93.

U.S. Department of Health and Human Services, Public Health Service, *Guidelines for the Conduct of Research in the Intramural Research Program at the National Institutes of Health.* Washington D.C. Oct. 1994.

CHAPTER FIVE

Responsible Laboratory Procedures

> I fully subscribe to the judgement of those writers who maintain that of all the differences between man and the lower animals, the moral sense or conscience is by far the most important.
>
> *Charles Darwin, The Descent of Man*

Responsible Practice

RESPONSIBLE laboratory procedures are crucial to the collection of research data as well as to the health and safety of researchers. Every laboratory routinely employs certain techniques or protocols, such as specific cell-culturing techniques (e.g., cell transfers, dilutions, cell counting, media preparation), irradiation techniques, tumor implantation procedures, neurologic examinations, animal anesthesis, electrophoresis procedures, and others. Documentation and citing of details regarding methodologies for specific experiments facilitate replication and aid in the generation of new data. Such documentation also serves as the final and absolute arbitrating reference when questions of technique arise.

Ultimately, the responsibility for health and safety within the laboratory rests with each individual there. However, the laboratory supervisor must set the standards of performance. It is the responsibility of the laboratory supervisor to make clear by example or by direction that the careless practice of laboratory procedures is no more acceptable than careless and sloppy science. The supervisor is responsible for establishing safe work practices and ensuring that trainees are informed about and fully understand any risks associated with the research. This is not to say that all experimentation must be totally free of risk, because this goal is unattainable, but it does mean that all reasonable and practical steps must be taken to minimize risks. The equipment must be maintained in good repair and be adequately designed to work properly. Individuals must be fully trained in the procedures they are to carry out, and research must be carefully analyzed to foresee potential accidents or failures. Contingency plans must be developed to meet at least the most likely emergencies.

The failure of institutions to establish safety procedures and guidelines is the equivalent of willful negligence. Employees, on the other hand, have an equal responsibility to adhere to safety policies that have been established—to be knowledgeable about safe laboratory practices and not diverge from these practices when they seem too time consuming, too much trouble, or not convenient at the moment. There are a number of reports and studies on laboratory safety which indicate that many exposures to pathogens and consequent infections occur not as the result of overt accidents but during the performance of routine procedures (National Research Council, p. 18). If trainees are uncertain of the proper procedures, they must be allowed the opportunity to seek clarification. This may be extremely difficult to ensure unless the laboratory supervisor creates a responsible learning environment—one in which the training experience reflects the highest standards of research integrity. The following guidelines and policies exemplify standards of responsible laboratory practices. However, they are not intended as a complete or final listing of potential hazards or of safe practices. Because of the diverse nature of research in university laboratories, additional procedures or information may be necessary.

Biosafety

- Do not mouth pipette.
- Manipulate infectious fluids carefully to avoid spills and the production of aerosol droplets.
- Restrict the use of needles and syringes to those procedures for which there are no alternatives; use needles, syringes, and other "sharps" carefully to avoid self-inoculation; and dispose of "sharps" in leak- and puncture-resistant containers.
- Use protective laboratory coats and gloves.
- Wash hands following all laboratory activities, following the removal of gloves, and immediately following contact with infectious materials.

- Decontaminate work surfaces before and after use and immediately after spills.
- Do not eat, drink, store food, or smoke in the laboratory.

Chemical Safety

Trainees should be knowledgeable about the regulations of their institutions with regard to the following:
- Transporting chemicals
- Operation of chemical hoods
- Flammable liquids: handling, storage, flash points, boiling points, ignition temperatures, sizes of containers
- Maximum quantities
- Highly reactive chemicals and explosives: reactivity, inflammability, toxicity, peroxidizable materials, incompatibles
- Toxic substances: inhalation, ingestion, contact with skin or eyes, toxicity, classes, corrosives, strong acids, strong bases, dehydrating agents, oxidizing agents, inhalation, use and storage
- Metals: safe use and storage, metal powders
- Chemical spills: clean-up procedures, quantity of spill, chemical and physical properties, hazardous properties of material, protective equipment needed
- Spill kits, emergency procedures (National Research Council, p. 19).

Principles of Safe Practice

Each laboratory should develop or adopt an operations manual that identifies hazards that will or may be encountered, and that specifies practices and procedures designed to minimize or eliminate risks. Personnel should be advised of special hazards, and should be required to read and follow the required practices and procedures. A scientist trained and knowledgeable in appropriate laboratory techniques and safety procedures and hazards associated with handling infectious agents must direct laboratory activities (U.S. Department of Health, p. 6).

Containment

The term *containment* is used in describing safe methods for managing infectious agents in the laboratory environment where they are handled or maintained. The purpose of containment is to reduce or eliminate exposure of laboratory personnel, other individuals, and the outside environment to potentially hazardous agents. *Primary* containment, the protection of personnel and the immediate laboratory environment from exposure to infectious agents, is provided by good microbiological technique and the use of appropriate safety equipment. The use of vaccines may also provide an increased level of personal protection. *Secondary* containment, protection of the environment external to the laboratory from exposure to infectious materials, is provided by a combination of facility design and operational practices. Therefore, the three elements of containment comprise: laboratory practice and technique, safety equipment, and facility design. The *risk assessment* of work to be done with a specific agent will determine the appropriate combination of these elements (U.S. Department of Health, p. 6).

Radiation Safety

Research with radioactive materials entails special procedures that must be in place before permits for use can be obtained. Authorization to use radioactive materials may be granted to any researcher who successfully completes the application, training, and review process. The review process ensures that the application form is complete, and that required training has been carried out; this means that responses to an examination were appropriate and adequate facilities are present. Facility requirements may include: (1) sufficient space for the number of personnel, (2) adequate ventilation, (3) impervious work surfaces, and floors in excellent condition, (4) a properly functioning fume hood, and (5) adequate equipment for the requested experiments, level of activity, radiation anticipated, and energy released. This equipment may include appropriate radiation detection equipment for the type of radiation and energy in question, shielding sufficient to reduce radiation levels, decontamination equipment, protective gear, remote-handling equipment, and hand protection. In addition, cautionary signs must appear on doors, waste disposal bins, disposal sinks, and areas where radioactive materials are present, and there must be double-locking mechanisms to secure inventories. All personnel working with radioactive materials are required to comply with safe-handling practices such as the following:

- Smoking, eating, application of cosmetics, and storage of food are prohibited in a laboratory where radioactive material is used or stored.
- Personal belongings shall be kept segregated from use areas.
- Designated mechanical pipettes shall be used; these shall be marked with "Caution—Radioactive Materials" tape or stickers.

- Work with radioactive material should be done expediently, but with care. The principle of [optimal] time, distance, and shielding should be used to minimize exposure.
- Work surfaces shall be covered with absorbent paper with the paper side up, plastic side down, to confine spills and aid in minimizing contamination; cautionary tape shall be used to affix the paper to the bench top.
- Disposable gloves shall be used, and the use of double gloves is recommended. Always remove the outer gloves after using a stock vial; then put on a new pair of gloves and proceed with the experiment. Remove protective apparel, wash and monitor hands prior to leaving the laboratory.
- Shield all stored sources of radiation.
- Perform surveys and do "wipe tests" of equipment and work areas after each use.
- All spills must be cleaned up and reported immediately (University of South Florida).

The case study in this chapter illustrates that responsibility for laboratory safety is a comprehensive responsibility. Researchers must recognize that trainees begin their laboratory work with very little formal training in laboratory safety, and that they generally view safety as secondary to their overall research goals. Therefore, formal safety programs that are offered early in a trainee's career can go a long way toward fostering responsible conduct.

The Use and Care of Laboratory Animals

The use of animals in research is currently a subject of controversy and disagreement. The debate is divided between those who feel that animals should never be used for research and those who feel using animals is not only justifiable but absolutely necessary for the advancement of medical research and the treatment of disease. The moral arguments for and against the use of animals in research focus on three ethical positions: 1. The *deontological*, or rights-based, position argues that all animals have specific rights that should be upheld. This is held to be true even though animals are dependent on others to assert those rights. 2. The *utilitarian* view determines the morality of specific research projects according to the balance between *harm* to the experimental animals and anticipated *benefits* to human beings. The determination of *risks versus benefits* is problematic because of uncertainty in assessing the positive and negative experiences of animals in relation to those of human beings, and in assessing the ultimate benefit of the knowledge to be gained. 3. The *humane treatment* or *communitarianism* position argues for improving the quality of life for laboratory animals without compromising the benefits that can be derived from their use in research. Thus, the animals' interest in avoiding pain, suffering, and distress is of enough moral significance to require that the research must advance the welfare of humans (Dresser).

Whatever moral position one assumes, there is general agreement among those who advocate the use of animals in research that there must be appropriate scientific support to justify experiments using animals as well as the number of animals to be studied. This is accomplished at research institutions through a mechanism known as the Institutional Animal Care and Use Committee (IACUC). The IACUC reviews proposed projects involving animals to ensure that they meet certain scientific and humane standards. The standards themselves are articulated in part by the following principles of humane research on animals as defined by the U.S. Public Health Service (PHS) and the U.S. Department of Agriculture.

- Procedures with animals will avoid or minimize discomfort, distress, and pain to the animals, consistent with sound research design.
- Procedures that may cause more than momentary or slight pain or distress to the animals will be performed with appropriate sedation, analgesia, or anesthesia, unless the procedure is justified for scientific reasons in writing by the investigator.
- Animals that could otherwise experience severe or chronic pain or distress that cannot be relieved will be painlessly sacrificed at the end of the procedure or, if appropriate, during the procedure.
- The living conditions of animals will be appropriate for their species and contribute to their health and comfort. The housing, feeding and non-medical care of the animals will be directed by a veterinarian or other scientist trained and experienced in the proper care, handling and use of the species being maintained or studied.
- Medical care for animals will be available and provided as necessary by a qualified veterinarian.
- Personnel conducting procedures on the species being maintained or studied will be appropriately qualified and trained in those procedures.
- Methods of euthanasia used will be consistent with the recommendations of the American Veterinary Medical Association (AVMA) Panel on Euthanasia, unless a deviation is justified for scientific reasons in writing by the investigator (National Institutes of Health).

It is the members of society at large who will ultimately determine the appropriate bounds for animal use in experimental research. In the meantime, all researchers must share the responsibility for ensuring the humane treatment of laboratory animals.

CASE STUDY 5

A Chronology of Contamination

Before beginning a radioactive phosphorus experiment—her first at the clinic—Shubha Murthy placed an absorbent pad on the lab counter. It was intended to soak up any spills. Unfortunately, she laid the material down with the laminated or non-absorbent side up, rendering it useless.

May 16, 1991 - 11:30 a.m.

Shubha Murthy, a Cleveland Clinic researcher, spills 55 millicuries of radioactive phosphorus inside laboratory NC 1-140.

May 16, 1991 - 1:30 p.m.

Murthy goes home, contaminating her car and home.

May 16, 1991 - 2:30 p.m.

Clinic employee Michael Salvatore inadvertently contaminates himself and then drives to the Cleveland Music School Settlement to practice piano. He contaminates two piano keyboards, a piano bench, the arm of a chair, two door knobs, the room key, and sheet music.

May 16, 1991 - 3:00 p.m.

A lab worker notices that a radiation detection device shows contamination. Paul L. Fox, who is in charge of the lab, advises staff not to report the spill immediately.

May 16, 1991 - 4:45 p.m.

Murthy reports what has happened to Radiation Safety Officer Steven J. Aron, Jr.

May 16, 1991 - 7:00 p.m.

Aron goes home for the day, assuming that the lab and all workers have been decontaminated. Ten people have been contaminated.

May 16, 1991 - 10:30 p.m.

Aron is informed by Clinic security that Salvatore's car is contaminated. Aron tells security to lock it, and that they will decontaminate it the next day.

May 17, 1991 7:00 a.m.

Clinic employees discover contamination on a drinking fountain in the building.

May 17, 1991 - 9:00 a.m.

Clinic employees discover that the sink on the north end of the building is contaminated, together with table, newspapers in the lounge, and the floors of several corridors.

May 17, 1991 - 3:00 p.m.

Aron, Salvatore, and another employee arrive at the music school to notify school officials that areas of the school may be contaminated. All areas are decontaminated.

May 18, 1991 - 1:00 p.m.

Acting on a tip, Nuclear Regulatory Commission (NRC) inspector James Cameron arrives at the Clinic and discovers contamination on a sink and laboratory door knob.

May 18, 1991 - 4:31 p.m.

Aron formally notifies NRC of the spill.

July 29, 1991

Aron is fired by Clinic officials. NRC issues a $7,500 fine for radiation safety violations.

Source: Nuclear Regulatory Commission Newsletter

Questions for Discussion

1. What are the elements that contribute to a safe laboratory environment?
2. What would you do if you witnessed a spill that resulted in contamination?
3. Is it appropriate to use an animal model if an alternative, nonliving system works?
4. Is repeated experimentation on a single animal justified if it means sacrificing fewer animals overall?

Recommended Reading

A.K. Furr (ed.), *Handbook of Laboratory Safety*, 3rd ed. CRC Press, Boca Raton 1990.

J.Y. Richmond, R.W. McKinney (eds.), U.S. Department of Health and Human Services, *Biosafety in Microbiological and Biomedical Laboratories*, 3rd ed. U.S. Government Printing Office, Washington D.C. 1993.

P. Singer, *Animal Liberation: A New Ethic for Our Treatment of Animals*.Random, New York 1975 (revised 1990).

J.A. Smith, K.M. Boyd, *Lives in the Balance: The Ethics of Using Animals in Biomedical Research*. Oxford University Press, Oxford 1991.

Works Cited

R. Dresser, *Animal Experimentation Biolaw: A Legal and Ethical Report on Medicine, Health Care, and Bioengineering*. University Publications of America 1986.

National Institutes of Health, Office for the Protection from Research Risks, *Public Health Service Policy on Humane Care and Use of Laboratory Animals*. Bethesda 1986.

National Research Council, *Biosafety in the Laboratory: Prudent Practices for Handling and Disposal of Infectious Materials*. National Academy Press, Washington D.C.

University of South Florida, *Users' Manual of the Safe Use of Radioactive Materials*. Tampa 1995.

U.S. Department of Health and Human Services, Public Health Service, *Biosafety in Microbiological and Biomedical Laboratories*. U.S. Government Printing Office, Washington D.C. 1993.

CHAPTER SIX

Responsible Mentoring, Training, and Supervision

> The place to improve the world is first in one's own heart and head and hands, and then work outward from there.
>
> *Robert Pirsig, Zen and the Art of Motorcycle Maintenance*

Mentorship—Obligations and Expectations

TRAINING and supervision are the mechanisms by which knowledge and skills are from one generation of researchers passed to the next. Mentors have an obligation as part of their professional responsibilities to participate in the process by providing training experiences for young scientists. A mentor is a person who is directly responsible for the professional development of a trainee (Guston, ch. 3 p. 51). Professional development includes both technical training, such as instruction in the methods of scientific research (e.g., research design, instrument use, and selection of research questions and data) and socialization in basic research practices (e.g., authorship practices and the sharing of research data). A mentor is *not* merely an advisor, nor is a mentor a supervisor who oversees the student's dissertation, or a role model who can influence a trainee indirectly, unknowingly, or from a distance (Guston, ch. 3 p. 52). A mentor's role is unique, more complex, and more crucial to the overall development of the trainee than that of the advisor, supervisor, or role model.

Many universities have written guidelines for the supervision or mentorship of trainees as part of their institutional research policies. These guidelines affirm the need for regular and personal mentor–trainee interaction. In addition, they suggest that mentors may need to limit the size of their laboratories to enable them to interact directly and frequently with all individuals under their supervision. Although there are many ways to ensure responsible mentorship, methods that provide continuous feedback, whether through formal or informal mechanisms, are apt to be the most successful (Guston, ch. 2 p. 65).

In *Guidelines for the Conduct of Research in the Intramural Research Program,* NIH characterizes the importance of and defines the standards for creating a successful mentoring experience:

> Research training is a complex process, the central aspect of which is an extended period of research carried out under the supervision of an experienced scientific mentor. This supervised research experience represents not merely performance of tasks assigned by the supervisor, but rather a process wherein the trainee takes on an increasingly independent role in the choice of research projects, development of hypotheses and the performance of the work. Indeed, if training is to prepare a young scientist for a successful career as a research investigator, it should provide the trainee with the aforementioned skills and experiences. It is particularly critical that the mentor recognize that the trainee is not simply an additional laboratory worker. Each trainee should have a designated primary scientific mentor. It is the responsibility of this mentor to provide a research environment in which the trainee has the opportunity to acquire both conceptual and technical skills of the field. In this setting, the trainee should undertake a significant piece of research, chosen usually as the result of discussions between the mentor and the trainee, which has the potential to yield new knowledge of importance in that field. The mentor should supervise the trainee's progress closely and interact personally with the trainee on a regular basis to make the training experience meaningful. Mentors should limit the number of trainees in their laboratory to the number for whom they can provide an appropriate experience. There are certain specific aspects of the mentor–trainee relationship that deserve emphasis. First, mentors should be particularly diligent not to involve trainees in research activities that do not provide meaningful training experiences. Second, training should impart to the trainee appropriate standards of scientific conduct by instruction and by example. Third, mentors should provide trainees with realistic appraisals of their performance and with advice about career development and opportunities (U. S. Department of Health, p. 4).

The Ethical Use of Trainees—The Laboratory Experience

In the highly competitive world of the research laboratory, the misuse of trainees through benign neglect may be a prevalent although unintentional practice (Guston, ch. 3 p. 57). Research advisors *do harm* to trainees if they use them in the laboratory as pairs of hands (Amundson, p. 160). Because of the unique relationship between mentors and trainees, it is important that the laboratory experience be mutually rewarding and supportive (Guston, ch. 2 p. 59). It is the responsibility of the mentor to ensure that the trainee's research is completed in a *sound, honest, and timely manner* (Guston, ch. 3 p. 59). "The ideal mentor challenges the trainee, spurs the trainee to higher scientific achievement, and helps socialize the trainee into the community of scientists by demonstrating and discussing methods and practices that are well understood" (Guston, ch. 3 p. 60).

In many research fields, concerns are being raised about how the increasing size and diverse composition of research groups affect the quality of the relationship between trainee and mentor (Guston, ch. 3 p. 60). As research laboratories expand, the quality of the training environment is at even greater risk. In 1989, the National Academy of Sciences (NAS), the National Academy of Engineering (NAE), and the Institute of Medicine (IOM) initiated a major study to examine issues related to scientific responsibility and the conduct of research. The Committee on Science, Engineering, and Public Policy convened a study panel to review factors that influence the integrity of science and the research process as it is carried out in the United States. As a result of this examination, the study panel proposed a framework of issues to consider for encouraging responsible research practices (Guston, ch. 2 p. 60). Those issues that relate to mentorship and the use of trainees were identified as:

- Assignment of mentors to students,
- Availability of mentors and appropriate forms of supervision,
- Degree of independence and responsibility for students and postdoctoral trainees,
- Types of duties assignable to students by mentors and supervisors, and
- Appraisals and communication of student and trainee performance.

In addition, the panel proposed the following steps to encourage responsible mentoring and supervision practices: (1) Each trainee should have a designated primary scientific mentor. (2) The preceptor (mentor) should provide each new investigator (whether student, postdoctoral fellow, or junior faculty member) with applicable government and institutional requirements for conduct of studies involving healthy volunteers or patients, animals, radioactive or other hazardous substances, and recombinant DNA. (3) The preceptor should supervise the design of experiments and the processes of acquiring, recording, examining, interpreting, and storing data. A preceptor who limits his/her role to the editing of manuscripts does not provide adequate supervision. Mentors have a professional obligation to pass along knowledge and skills to the next generation of scientists by example, as well as by instruction (Guston, ch. 2 p. 141).

Sexual and Other Harassment

Because there is an imbalance of power in the mentor–trainee relationship, serious problems may arise, such as sexual harassment, misunderstandings, envy from co-workers or spouses, and failure on the part of some male participants to treat a female mentor or trainee professionally. Problems are especially common in opposite-sex mentoring relationships (Rawles). Romantic liaisons between mentors and trainees, especially graduate students, are generally discouraged without being prohibited (Guston, ch. 3 p. 59). Although more and more educational institutions are moving toward establishing guidelines for acceptable and unacceptable behavior, defining sexual harassment is key to developing effective policies. Besides the most obvious offenses, more subtle behaviors such as telling off-color jokes, using derogatory names, displaying sexually revealing pictures, and making casual hand gestures, can all be interpreted as sexual harassment. Sexual harassment is more than a moral, legal, or financial-liability concern. It is a concern for protecting and maintaining an atmosphere consistent with academic ideals and the ethics of responsible conduct. In a condition of fear or emotional discomfort, academic goals cannot be achieved. In a research laboratory where mentors have direct supervision over trainees and are protected by tenure and academic freedom, it is their obligation to create an environment that promotes ideals and values, thereby diminishing the likelihood of sexual harassment (Riggs, Murrell, Cutting, p. 1).

In 1994, the National Association of Scholars (NAS) issued the following position statement on sexual harassment and academic freedom:

> It is the position of the NAS that sexual harassment is always contemptible, because it subverts education, and is particularly damaging in an academic setting. The National Association of Scholars believes that college and university authorities should respond to instances of sexual harassment promptly and firmly. Tenure does not protect members found guilty of coercing sexual favors. Nor is academic freedom an impediment to the effective enforcement of prohibitions against lesser forms of harassment, such as inappropriate touching, that also exploit individuals and undermine the educational process. Such behavior constitutes a serious violation of an educator's responsibility and is morally wrong. It cannot be tolerated. However, academic freedom and the rights of individuals can be—and have been—violated by misguided efforts to combat sexual harassment. Too many institutions have adopted vague definitions of harassment that may all too easily be applied to attitudes or even to a scholar's professional views. Not surprisingly, a chill has descended on academic discussions of sensitive but legitimate topics, such as human sexuality, sex differences, and sexual roles. Worse, procedures have been widely adopted that violate the canons of due process. In a number of cases, the reputations and careers of innocent persons have been severely damaged as a result of unwarranted actions taken by college and university authorities.

Specifically:

- The criteria for identifying "harassment" are often nebulous, allowing for expansive interpretations of its meaning. For example, at one major public university, the official definition stated that "Sexual harassment can be as blatant as a rape or as subtle as a look. Harassment . . . often consists of a callous insensitivity to the experience of women;" in another official publication the university further advised that sexual harassment "is broadly defined to include behavior that may not be considered overtly sexual." Language like this can only breed confusion, resentment, and injustice.

- When definitions of sexual harassment are expanded to include opinions and attitudes, academic freedom is violated. Such definitions have already significantly inhibited discussion inside and outside the classroom. Ambiguous phrases like "callous insensitivity to the experience of women" have inspired complaints against professors accused of slighting gender-based literary analysis, or who have discussed theories and findings, such as Freud's, that run counter to the prevailing consensus about sexual differences.

- Some definitions of sexual harassment embrace a wholly subjective test of its occurrence—for example, the complainant having been made to "feel uncomfortable." Proof relies not on the objective behavior of the alleged harasser but on how one person perceived that behavior.

- Charges of sexual harassment are sometimes entertained long after the alleged offense, when the memories of the parties have faded, their motives have altered, and evidence has been lost.

- Mid-level administrators with meager academic experience but a strong commitment to fashionable causes are frequently accorded a major role in drawing up harassment regulations, interpreting them, counseling complainants, investigating charges, administering hearings, and determining guilt and penalties. Sometimes one and the same person performs all of these functions and, in addition, encourages students and others to make harassment charges. This leads to violations of academic due process.

- Investigations of alleged sexual harassment can provide a pretext for engaging in the ideological persecution of persons whose views are out of favor. When amorphous harassment regulations are enforced over-zealously, they encourage frivolous or self-serving charges from discontented or vindictive individuals.

- Collective penalties have been imposed on entire academic departments and groups of students for actions, not always proven, of a few individuals. At one prestigious private university, every student, faculty member, and staff member was ordered to attend a three-hour seminar on sexual harassment because a federal agency found fault with the university's judgement in a case involving a single alleged harasser. It need hardly be said that finding persons guilty by association, or making a whole category responsible for actions only committed by one or a few, violates the principle of individual responsibility basic to our society. It is likely to arouse resistance against efforts to combat genuine sexual harassment.

- Some of the sanctions now in force aim at thought reform, frequently through compulsory "sensitivity training" programs. These programs often operate to humiliate those suspected of holding "incorrect views." They also suggest that being white, male, or heterosexual constitutes a presumption of guilt. When required for faculty or students, these programs constitute an assault on individual dignity and freedom.

What we are witnessing, in short, is the transformation of a clear behavioral offense into a ubiquitous "thought crime", and the substitution of psychological manipulation for rational discussion. The resulting confusion of genuine harassment with less serious acts, and even with beliefs, brings anti-harassment policy into needless conflict with academic freedom. This confusion is bound to diminish the opprobrium that rightly attaches to sexual harassment. In the interest of protecting students and staff from genuine harassment and of preserving academic freedom, The NAS urges institutions of higher education to:

- Define sexual harassment precisely, confining it to individual behavior that is manifestly sexual and that clearly violates the rights of others,
- Set a reasonable statute of limitations on bringing sexual harassment charges,
- Separate the offices of investigator, prosecutor, judge, and jury; observe the requirements of due process to ensure the right of the accused to make an adequate defense,
- Punish those who knowingly lodge false accusations of harassment, and
- Act against proven harassers forcefully, by dismissal if necessary, instead of coercing opinions and restricting speech.

Anything less fails to protect students, faculty, and staff from sexual harassment. Anything more threatens individual freedoms and, most conspicuously, academic freedom (National Association).

Discrimination

Refusal to mentor or denial of the mentoring experience can constitute a form of gender and/or racial discrimination (Spector, p. 3). In a lawsuit filed by a female research psychiatrist at the National Institutes of Health, the complainant stated she was denied professional opportunities that her male colleagues had been afforded. Those opportunities included treating patients on long-term treatment trials, co-research with drug companies, and being involved with research in more valued biological studies. She also alleged being denied respectful supportive mentoring opportunities that were offered her male counterparts. A federal jury agreed that she had been discriminated against in the denial of mentoring, resulting in "damage to [her] professional reputation." Her attorney claimed that "this case is of major significance for women scientists and other professional women, because how one progresses in one's career depends to a large extent on mentoring. Someone with an established reputation can give you entree to all sorts of opportunities" (Spector, p. 11). Whatever the long-term legal ramifications of this case may be, one of its early effects has been to reemphasize the importance of mentoring in the early stages of a scientific career. The creation of a hostile environment, whether by benign neglect or deliberate acts on the part of senior scientists, impedes the advancement of science. How trainees are treated during their early training experiences affects not only the course of their careers and their roles as future mentors, but also the integrity of the research process.

CASE STUDY 6

A Trainee's Dilemma

Laurie has recently been accepted into a graduate program in marine biology at a large university. She is looking forward to her graduate studies in this field, and she appreciates the fact that this university has an outstanding marine biology program. At the initial meeting with the professor she is considering as a mentor, she learns that she is the only woman in the program. Throughout the interview the professor acts very paternal toward her; he addresses her as "honey", and at one point pats her on the head. Laurie is aware that he has an outstanding reputation as a leading expert in marine life, and that to have him as her mentor would be extremely beneficial to her career advancement. However, the interview leaves her with an uneasy feeling about how she will be treated as a trainee.

Questions for Discussion

1. What do you consider inappropriate behaviors in a research laboratory?
2. What information about a mentor would you consider important for you to know prior to accepting a mentorship?
3. What behaviors would you characterize as harassment?
4. What attitudes would you characterize as racial or gender discrimination?

Recommended Reading

K. Bursick, *Perceptions of Sexual Harassment in an Academic Context*. Sex Roles 27:7/8 (1992), pp. 401–412.

T.A. Krulwich, P. J. Friedman, *Integrity in the Education of Researchers*. Academic Medicine (Supplement) 68:9 (1993) S14–S18.

D.C. Locke, *Increasing Multicultural Understanding: A Comprehensive Model*. Sage, Newbury Park 1992.

G.T. Perkoff, *To Be A Mentor*. Educational Research and Methods 24:8 (1992) pp. 584–585.

Trying to Change the Face of Science. Science 262:5136, 12 Nov. 1993, pp. 1089–1136.

Works Cited

N.R. Amundson, *American University Graduate Work*. Chemical Engineering Education 21:4 (1987) pp. 160–163.

D.H. Guston, *Scientific Principles and Research Practices*. Chap. 2 In *Responsible Science: Ensuring the Integrity of the Research Process*, vol 1. National Academy Press, Washington D.C. 1992.

D.H. Guston, *Mentorship and the Research Training Experience*. Chap. 3 In *Responsible Science: Ensuring the Integrity of the Research Process*, vol 2. National Academy Press, Washington D.C. 1993.

National Association of Scholars, *The Chronicle of Higher Education*. 26 Jan. 1994.

J. Rawles, *The Influence of a Mentor on the Level of Self-Actualization of American Scientists*. Thesis, Ohio State University 1980.

R.O. Riggs, P.H. Murrell, J.A.C. Cutting, *Sexual Harassment in Higher Education: From Conflict to Community*. ASHE ERIC Higher Education Report 93:2 (1994), pp. 1–4.

B. Spector, *Jensvold Verdict*. The Scientist, 16 May 1994, pp. 3–11.

U.S. Department of Health and Human Services, Public Health Service, *Guidelines for the Conduct of Research in the Intramural Research Program at the National Institutes of Health*. Oct. 1994.

CHAPTER SEVEN

Research Grants and the Review Process

Richard B. Streeter

> Each of us should lay aside all other learning to study
> only how he may discover one who can give him the
> knowledge enabling him to distinguish the good life
> from the evil.
>
> *Plato, The Republic*

ANY QUESTIONS regarding institutional and individual ethics surround the review process for research grant proposals. The most basic issue is the consonance of the proposed project with the missions of the researcher's institution and the sponsoring agency. To determine this matter, grant proposals are subjected to internal review by the researcher's institution and to external peer review by the proposed sponsor. This review process does not end with the award of a grant. Rather, the review process is an ongoing one which involves the researcher and the university, as well as the researcher and the sponsor.

Internal Institutional Review

The academic researcher is usually part of a department within a college of the university. Therefore, the prudent central research office will make certain that the proposed project has been approved by the researcher's department and college before continuing the review with respect to other required institutional policies and procedures. The project review begins with these two units' approving the project for consonance with the institutional mission.

The department and the college must also evaluate a researcher's ability to conduct the project and at the same time continue to perform other duties, such as teaching, serving on committees, advising students, and participating in other research grants. Many times, the department will conduct a peer review of the scientific merit of the proposal. In some cases, the proposed project may involve additional compensation for personnel, commitments for space and equipment, and matching funds. If the necessary components for the success of the project are not available, then the department/college must assess whether these resources can be obtained, or whether they can be procured via the grant. The institution always needs to ensure that a project can be completed before any proposal is submitted.

Once these assurances have been given in writing, the central research office needs to determine if the proposal is in compliance with institutional and sponsor policy and guidelines. This review can involve such considerations as intellectual property; conflict of interest; proposed subcontracts and/or consultants; use of human subjects, animal subjects, or biohazardous or radioactive materials; accuracy of the budget; and proposal format. Any of these items could create an ethical problem for the institution and/or the researcher.

Each of the items in question needs to be evaluated separately. Intellectual property considerations vary greatly according to the sponsor. If the proposal is to be submitted to a private-sector sponsor (IBM, for example) ownership rights will be subject to negotiations. If the sponsor is to be the federal government, then the rights are predetermined by public law. In any event, the university needs to determine the propriety of the information submitted, and label it accordingly. The ownership of intellectual property may have an impact on the publication rights of the university and the researcher.

Currently, the federal government requires an assessment of any potential conflicts of interest. The National Science Foundation has issued policy guidelines, and the National Institutes of Health has issued procedural guidelines. Both agencies have defined conflict of interest. The researcher needs to disclose potential conflicts, and the institution needs to assess potential conflicts to determine if the project can be structured so that these can be avoided or obviated. The institution must also certify that it has appropriate policies and procedures in place, and that the proposal has been reviewed in accordance with those procedures.

Subcontractors and consultants need to be identified in the proposal so that the sponsor can evaluate their potential contributions to the project. Although prior agency approval is implied when project is funded, the institution needs to ensure that there are proper review criteria in place, including screening for potential federal debarment, to prevent concerns regarding favoritism during the process. The level of a subcontract may also be problematic. While a small subcontract may cause concern about the openness of competition, a large subcontract may raise other questions. Normally, a subcontract should be less than 51 percent of the proposal as a whohle. In instances in which the subcontractor has a larger portion, an institution may be criticized for assisting the sponsor in avoiding a competitive bid process by passing funds to a subcontractor who may be ineligible for the prime contract. Both small and large subcontracts can be "sanitized" by adequate justification in the proposal and/or a competitive bid process.

Nowhere are ethical concerns more valid than in proposals that involve human or animal subjects. All institutions using human subjects must have an Institutional Review Board (IRB), or must seek the review of an independently contracted IRB. The NIH allows sixty days from the submission date of a proposal for IRB approval of the research protocol. The central research office must have a process for tracking and certifying that such a review and approval have taken place. Similarly, the Institutional Animal Care and Use Committee (IACUC) must review, and the research office must track, all animal protocols for certification to the sponsoring agency. Both the IRB and the IACUC should be primarily concerned with the ethics of the protocol. Though a university may reject a protocol that has been approved by these boards, the institution may not approve a protocol that has been rejected. The authority of such boards to review ethical issues is guaranteed in the assurances that the institution negotiates with the Office for the Protection of Research Risks (OPRR) in NIH.

At many institutions, the committee that reviews biohazardous materials (blood, DNA, pathogens, etc.) also reviews and approves permits to allow researchers to use radioactive materials and/or controlled substances. Given the political sensitivity of these topics, it is incumbent on the researcher to make certain that the proposal receives approval before, or shortly after, the project is submitted to the sponsor. The potential for exposure of workers and students to these hazards presents an ongoing ethical and safety issue for the researcher.

During the preparation of a budget, the researcher and the central research office need to make certain that all reasonable costs have been anticipated and disclosed. If a project is under-budgeted, the work plan may not be completed, and commitments to the sponsor may not be honored. This issue may resurface during negotiation of an award. The institution and the researcher must be careful not to jeopardize the scope of the work by preparing an unrealistic budget merely to obtain the grant. The institution must maintain an ethical posture and reduce the scope of the work if the budget is not sufficient to complete the project as originally planned.

Though more clerical in nature, the format of a proposal may present some interesting ethical dilemmas as well. Most agencies have moved to page limits on proposals. The more astute ones have also specified type style and font size. In the age of word processing, the temptation to down-size fonts to meet page limitations is more and more of a problem.

A more recent ethical problem has developed as more researchers find it difficult to obtain full-time tenure-track faculty positions; who is then eligible to be a principal investigator on a project? Though most institutions leave this decision to the department/college, the institution would be well served to establish minimum criteria for an institutional principal investigator. Because the institution is legally responsible for the project, universities normally expect the principal investigator to have a contractual appointment and an employer–employee relationship with the university. The institution must be able to hold someone accountable for the project.

Once the internal review of a proposal has been completed, the institution will submit the project for external review by the proposed sponsor.

External Peer Review

Peer review is the mechanism by which agencies (and journals) assess the scholarly work of potential researchers (authors). Reviewers who are recognized experts in a given field are invited by the agency to evaluate proposals submitted by their colleagues (peers). The best examples of the peer-review process in federal sponsoring agencies are found at NIH and NSF. Both agencies have a long, established history of peer review.

The NIH has both internally and externally funded research projects. The external (extramural) proposals may fall into one of three categories— grants, contracts, or cooperative agreements. Though the review process differs for each of these three groups of proposals, the grant process is the one most commonly encountered at NIH.

The initial application is submitted by a researcher to the Division of Research Grants (DRG) in NIH. The DRG reviews the application for format requirements and relevance to the NIH mission, and assigns it to an institute and to a Scientific Review Group (SRG). This assignment is based on the legislative mandate of the institute and on the expertise of the SRG. If the subject matter crosses two institutes, the DRG may make a dual assignment.

The institute staff makes an initial judgement regarding the technical merit and the appropriateness of the research before proposals are sent to the SRG for review. During this process, the staff decides if any potential conflict of interest exists between the researcher and an SRG reviewer. Although reviewers are required to sign a certification of lack of conflict, the staff may still decide not to send a proposal to a particular SRG member.

SRG members are expected to review all the proposals sent to them; however, certain reviewers are designated as *primary* or *secondary* reviewers for specific proposals. The primary and secondary reviewers are expected to provide detailed written evaluations of their designated proposals based on the review criteria. Other reviewers are designated as *readers* or *discussants* for several proposals. The primary and secondary reviewers and the readers and discussants lead the discussions of their proposals when the SRG meets. If necessary, special experts may be requested by the reviewers to review certain proposals before the SRG meeting.

SRGs consider the following aspects of a proposal: The strengths and weaknesses of the proposal itself, the qualifications of the researcher, the resources and environment in which the project is to take place, the rationale for the budget, any overlap with the researcher's current and/or pending support, any possible misconduct in science that may be represented in the proposal, the protection of human and animal subjects, the adequacy of safeguards for the use of hazardous materials, and the inclusion of minorities and both genders as subjects. The SRG may decide that the proposal is meritorious (and thus assign it a priority score), recommend that the proposal not be considered further, or recommend that the proposal be deferred. A summary statement of the reviewers' comments, the priority score, and the recommendations are sent to the appropriate NIH council as well as to the researcher. These documents guide the council and the institute in deciding which applications to fund (U.S. Department of Health).

The National Science Foundation also has a well-developed and long-standing peer-review process. Though its process is less structured, the NSF does use expert reviewers and study sections. The study sections tend to correspond to the organizational structure of the Foundation; for example, Analytical and Surface Chemistry. Proposals are assigned to an organizational unit, which conducts a preliminary review before the proposals are sent to the reviewer. Though each unit has a list of reviewers, many of the reviews are accomplished by mail rather than through a formal study-section meeting. The staff, like that of NIH, attempts to remove any ethical conflicts from the process, and all NSF reviewers sign a conflict-of-interest statement.

Both the NIH and the NSF have common standards of performance that generally apply to all peer review. As previously mentioned, reviewers are asked to sign a certification which specifies that they have no conflict of interest in evaluating the proposed projects. Reviewers must also leave the room when an application from their own institution is being discussed. In addition, consultants are not permitted to participate in any discussion of a proposal in which they are to be a participant, and reviewers are not allowed to lobby on behalf of their own research. Throughout the review process it is important that the confidentiality of the application be protected.

The NSF review criteria are similar to those of the NIH—an assessment of the researcher's capabilities to perform the project, the institution's resources, anticipated contribution to the relevant field of science, application of the research to technological and/or societal problems, and the effect of the proposed study on the scientific infrastructure (e.g., graduate-student support). The following categories are used for ranking proposals:

- Excellent (roughly the top 10 percent)
- Very good (roughly the top 30 percent)
- Good (the middle 30 percent)
- Fair (the lowest 30 percent)
- Poor (should not be supported) (National Science Foundation).

Because most federal agencies use external peer review for evaluating research proposals, the major potential for ethical conflict occurs not within the peer-review process, but in the absence of such a process. The major manifestation of absence of peer review arises in the area of congressional "earmarking of funds." Because representatives to Congress represent the interests of their election districts, some argue that Congress has the right, if not the responsibility, to earmark funds to benefit local constituencies. Most researchers believe that earmarking federal funds for research constitutes a serious breach of ethics because it bypasses the very system that is supposed to validate the quality of the research—peer review. This is of paramount concern to faculty researchers because the number of earmarks, including those for research programs, has drastically increased in recent years. This increase has occurred at the very time when basic research funding is, at best, flat. If federal funding for basic research shrinks while earmarking increases, then the whole system of peer-reviewed research is endangered. One may well ask the question, "What is an ethical substitute for peer review?"

Rights and Responsibilities of the Researcher and the Institution

When a research grant is submitted, it is usually submitted officially by an institution; therefore, the grantee is the institution as a whole. The president of the institution is legally responsible for the grant on behalf of the institution. This relationship is typically well-defined in the grant instrument, which normally incorporates—either directly or by reference—several key elements:

- An award letter, which states the basic requirements and includes other attachments that apply
- An approved budget in varying levels of detail
- The proposal, included by reference, to which the relevant terms and conditions are appended
- The specifics of the award, which may differ depending on the type of award given to the institution

There are three types of award instruments: (1) a *contract*, which represents a procurement of service by the federal agency (usually solicited through a *request for proposal*); (2) a *cooperative agreement*, which is used when the performance of the project requires significant agency involvement, and (3) a *grant*. Most research projects to universities result in grant instruments that normally provide full project costs. In return, the institution makes a commitment to perform the project in a prudent manner and in accordance with the agency's provisions. When the grantee accepts the grant award, the project can begin.

To assure that institutions abide by the terms and conditions of a project, all federal grantees are required to develop internal procedures for conduct of grant projects. Though the researcher has the prime responsibility for technical and overall supervision of the project, the institution assumes ultimate responsibility for compliance and accountability. Compliance and accountability are manifested in a number of federally mandated regulations, many of which deal directly with the ethical conduct of the project.

Post-Award Compliance and Accountability

Though many of an institution's policies dealing with compliance and accountability are driven by federal mandates, they must also be consistent with the mission of the institution and its standard operating procedures. This latter element may sometimes cause more concern than the federal mandates.

Most institutions (private or public) have been in existence since long before research grants became an important part of the organization. Therefore, the researcher and the central research office may have to learn to conduct research grants in an environment that is predominantly geared toward another function. That function may or may not readily accommodate an influx of external, federally-funded projects. The researcher may thus be faced with the ethical dilemma of completing obligatory technical work *despite* institutional policies. To reduce the incidence of such dilemmas and aid the post-award process, it is particularly important that the institution review every proposal before it is submitted to the sponsor.

In many public institutions, federal research funds are deposited in the state treasury, which means that those funds become subject to the laws and administrative codes of the state, as well as to federal terms and conditions. In some cases state codes may be in conflict with federal conditions, or may be subject to additional restrictions. The researcher's task may become a complex one of balancing the rules of the institution against those of the sponsor.

The job of the central research office is also compounded when such conflicts exist. In fact, *there is an inherent potential ethical conflict throughout the whole post-award process*. On the one hand, the institution and the researcher want the project to succeed for the mutual benefit of both; on the other hand, the institution, as grantee, must also ensure the researcher's compliance and accountability.

One issue of accountability that has already been mentioned in this chapter is fiscal accountability. Most granting agencies have guidelines for the expenditure of grant funds; therefore, the institution must have systems in place to monitor the expenditure of these funds. Some budget categories require agency approval if the researcher wishes to transfer funds from one category to another. Although this burden has been greatly reduced for institutions that are part of the Federal Demonstration Project, there are still a number of matters that need sponsor approval—for example, foreign travel, change in effort of key personnel, and special-purpose equipment. Because the institution is legally responsible for cost overruns and financial irregularities, many institutions have established policies for dealing with financial misconduct.

The NIH/OPRR does not consider financial misconduct subject to the regulations regarding misconduct in science; therefore, institutions need to have policies and processes of their own to handle allegations of financial misconduct. One simple way of establishing such a policy is to use the federal scientific misconduct policy as a model. This at least ensures some measure at consistency. In the absence of a federal policy, institutions without an internal policy have in effect left the ethical issue of financial abuse on research grants unresolved.

An institution also needs to assure project reporting and timely execution of research projects. This involves an ethical balance between interfering with the technical conduct of the work and remaining in compliance with the federal sponsor's reporting requirements. Because numerous institutions have demonstrated a weakness in this area, both sponsors and auditors have targeted this issue for action. Sponsors such as the Department of Defense now withhold new awards if institutions are delinquent in project reporting. Auditors performing A-110 systems audits have chastised institutions for not developing a centralized tracking process for technical reports. The net result of untimely technical reporting is a negative impact on researchers, most of whom share no responsibility in the problem. The worst-case scenario involves a researcher who leaves the institution without completing the technical reporting requirements on completed projects, thereby placing the institution in a delinquent status and with no direct control over the former researcher.

The most important issue associated with compliance and accountability is the relationship between the central research office and the researcher. The role of the research office in these areas centers around enforcement, training, and audit. The researcher shares this role, however, because these activities usually involve technicians who report to the researcher; therefore, the researcher and the research office must be partners in the endeavor, even though the research office may sometimes be forced to close a researcher's lab for noncompliance.

In recent years Congress and federal sponsors have dramatically increased compliance requirements in an area best described as *public policy issues*. These issues include civil rights, environmental impact, elimination of architectural barriers and other disability concerns, use of human subjects, abortion and use of fetal tissue, biosafety, animal welfare, the drug-free workplace, debarment and suspension of contractors, lobbying, data rights, intellectual property rights, scientific misconduct, and most recently, conflict of interest. Though all of these issues present ethical concerns, biosafety and the use of human subjects, animals, radioactive materials, and controlled substances remain among the most politically and emotionally sensitive.

The use of human and animal subjects, as previously discussed, is governed by an assurance of compliance negotiated between the institution and NIH/OPRR. The most difficult area to monitor is that of human subject research, because it generally occurs in the field and not in a laboratory. Use of radioactive materials, biohazardous materials, and controlled substances is usually contingent on a license from a state or federal agency. Documents of this type require the institution to enforce certain negotiated terms and conditions. Therefore, most research offices have a staff that inspects affected labs for compliance, or actually operates the vivariums. Infractions are referred to the appropriate faculty committee, which may impose sanctions or require preceptorships or the closing of laboratories.

As in the case of financial misconduct, federal scientific misconduct requirements do not apply to human and animal subjects or biohazard protocols. Therefore, each institution needs to provide a procedure by which misconduct may be reported and investigated. Normally, a Biohazard Review Committee is responsible for radiation safety, biological and hazardous material safety, and monitoring the use of controlled substances. An Institutional Animal Care and Use Committee regulates animal use in research and/or exhibition on campus. An Institutional Review Board regulates the use of human subjects in research. Misconduct may include, but need not be limited to, deviations from approved protocols, falsification of data, questionable practices, inhumane treatment of human or animal subjects, violation of state or federal regulations, and misuse or illegal distribution of controlled substances. Allegations of misconduct are reported to the chair of the appropriate committee.

In recent years, federal and state agencies have begun to emphasize the training of faculty and staff who deal with these issues. The U.S. Nuclear Regulatory Commission and the NIH/OPRR have for years required at least some training in proper research, but the scope and quality of the training and the range of categories of trainees (from students to faculty) have been greatly expanded. Therefore, research offices and researchers must work together to provide needed courses or mentorships to comply with the new training requirements.

Another recently expanded area is the OPRR requirement for field audit. Though biohazard and animal regulations have required inventory-control audits for a number of years, and all compliance committees are charged with continuous monitoring of approved protocols, the OPRR has only recently required assurances that include field audits of protocols. This requirement can be reasonably satisfied by having research-office staff conduct field audits of informed-consent documents. Just as the IRBs review protocols and require changes in the approved informed-consent form, the OPRR insists that the IRB (or research-office staff) validate that the most current approved form is in fact being used. These audits also verify that informed consent is being obtained by IRB-approved research staff.

The underlying ethical themes that pervade research review are conflicting roles and relationships. The federal sponsor, the researcher, and the institution are *all* required to be research advocates as well as research policers. The sponsor provides peer review, funding, and in some cases, technical assistance. In return, the sponsor expects the institution and the researcher to provide technical results *and* compliance and accountability. Though the federal sponsor depends on other audit agencies to conduct oversight for compliance and accountability, the institution and the researcher are responsible for carrying out these functions as they perform the research. Therefore, the institution and the researcher always find themselves in the dual role of policer and policee. How well they manage their relationship determines the amount of ethical conflict they will experience.

CASE STUDY 7

Who is in Charge Here?

Occasionally, a project comes along that gives the central research office faith in Murphy's Law. Dr. X, an hourly employee in an institute, received permission from his dean and director to put together an interdisciplinary proposal that would involve six colleges and eight departments. Dr. X purportedly had extensive experience in developing this kind of project, and would need minimal only help from the research office.

The sponsoring agency had not previously issued a request for proposals to the university community, had no experience with peer review, and had none of their own proposal or budget forms. The fact that this was a new program made matters worse, because the sponsor had no previous program guidelines to work from.

The project under consideration was a comprehensive program for bringing university resources to bear on urban problems. The university had an extensive track record in providing service and technical assistance to local governments, but most of those interactions had occured in a consulting relationship, or as training courses that had already been defined by the client.

Dr. X, knowing that the topic was of interest to the new university president, requested that his dean set up a meeting of chairs, deans, research-office staff, government-relations office staff, and the president. At the meeting the president offered support for the project, immediately making it a high-profile proposal. It was against this background that the internal-review process began.

Internal review

Weeks after the previously referenced meeting, little progress had been made on the proposal itself. It became apparent that Dr. X had always had staff prepare his proposals, and had no idea of how or where to start the effort on his own. His director had experience only in single-researcher training projects, and his dean was offering only moral support. Because the project was so complex, and had become a high-profile issue, the research office offered assistance well beyond its normal role.

A research-office staff member was assigned to the budget, and the director of the office assisted Dr. X in outlining the steps that had to be taken to assemble the material needed to write the text of the RFP. The "final" draft was reviewed by a consultant provided by the research office. After the review, the consultant met with Dr. X and his director and suggested that the entire text be reorganized to more specifically address the issues discussed in the RFP. In the meantime, no work could proceed on the budget because the key personnel had not been identified, and the work plan had to be redone. On the day before the agency deadline, the proposal was completed and submitted; however, the internal form used to verify university commitments had not been completed. (The final version of this form was not completed until six months later.)

The entire experience raised several ethical questions that still needed to be resolved: (1) If the project was so important, why did the dean's office fail to provide the necessary support? (2) How valid were the commitments for money and faculty time in the budget? (3) Why was Dr. X, an hourly employee, given the task of preparing the proposal? (4) Who was the real principal investigator (PI)? (Dr. X wrote the proposal, but, as an hourly employee, he was ineligible to be the PI. So, on submission, his director was listed as the PI.)

The peer review

The funding agency had no experience in peer review and, during the process, misinterpreted the forms, which they had borrowed from another agency. As a result, they offered to award the university only half the funding originally requested for the project. When Dr. X and his director were contacted by the agency, they were understandably upset. The agency, when informed of the problem by Dr. X, said that the university's research office had filled out the forms incorrectly. Dr. X and his director informed their dean of the research office's "mistake." Their dean immediately wrote an accusatory letter to the vice president for research who, in turn, chastised the director of the research office for his staff's error.

The director of the research office reviewed the proposal with his staff and informed all university "combatants" that the proposal and the forms had been completed correctly. He then called the agency and discussed the matter with the program officer, who subsequently admitted his agency's error. However, the funding had already been awarded, and there were no funds left to correct the error. The research office suggested that the agency seek permission to borrow funding from the subsequent year's appropriations or from year-end reserves. Eventually, the agency agreed to add the necessary funding if the university would submit a request in writing that acknowledged *their* (the university's) error in the proposal submission. The vice president for research reluctantly wrote the requested letter, and the funding was awarded.

Several ethical issues were raised during this part of the case: (1) Given that the university was the grantee, why did the agency notify Dr. X? (2) If the agency was going to notify anyone other than the research office, why did it fail to notify the PI? (3) Since the error was that of the agency, why did the agency insist that the university admit to being at fault? (4) Why did the university admit to being at fault when it was not?

Questions for Discussion

1. Who should be permitted to be a researcher, and why should the institution be concerned with this issue?
2. What constitutes a potential conflict of interest for an agency reviewer?
3. If earmarking is abhorrent to the peer review process, then why do institutions accept or seek such funding?

4. If the institution is the legal recipient of a grant award, how much control should it exercise over the principal investigator?
5. Who should be held responsible if a researcher overspends his/her budget?
6. When grant rules and institutional rules conflict, what is the appropriate role of the research office? Of the researcher? Of the institution?

Recommended Reading

D.E. Chubin, *Peerless Science: Peer Review and U.S. Science Policy.* State University of New York, Albany 1990.

H.-D. Daniel, *Guardians of Science: Fairness and Reliability of Peer Review.* VCH, Weinheim 1993.

K.S. Shrader-Frechette, *Ethics of Scientific Research.* Rowman and Littlefield, Lanham 1994.

U.S. Congress; House Committee on Science, Space, and Technology, Subcommittee on Investigations and Oversight, *Maintaining the Integrity of Scientific Research.* Hearing before the Subcommittee on Investigations and Oversight of the Committee on Science, Space, and Technology, U.S. House of Representatives, One Hundred First Congress, first session, June 28, 1989. U.S. Government Printing Office, Washington D.C. 1990.

U.S. General Accounting Office, *Peer Review: Reforms Needed to Ensure Fairness in Federal Agency Grant Selection.* Report to the Chairman, Committee on Governmental Affairs, U.S. Senate, Washington, D.C. 1994.

Works Cited

National Science Foundation, *Proposal Evaluation Form.* Washington, D.C. 1990.

U.S. Department of Health and Human Services, Public Health Service, National Institutes of Health, *Orientation Handbook for Members of Scientific Review Groups.* Bethesda 1992.

Works Consulted

9 CFR Chapter 1, *Animal Welfare Act.* U.S. Government Printing Office, Washington D.C. 1994.

The DEA: 21 CFR 1300 to 1316.99. U.S. Government Printing Office, Washington D.C. 1994.

Florida Department of Health and Rehabilitative Services, *Florida Administrative Code,* Chapter 10D-91. Tallahassee 1994.

National Science Foundation, *Grant Policy Manual*. Washington D.C. 1989.

45 CFR Part 46, *Protection of Human Subjects*. U.S. Government Printing Office, Washington D.C. 1991.

U.S. Department of Health and Human Services, Public Health Service, Center for Disease Control, *Biosafety in Microbiological and Biomedical Laboratories*, 3rd ed. Washington D.C. 1993.

U.S. Department of Health and Human Services, Public Health Service, *PHS Grants Policy Statement*. Washington D.C. 1994.

CHAPTER EIGHT

Intellectual Property

Lawrence R. Oremland

> The last temptation is the greatest treason: To do the
> right deed for the wrong reason.
>
> *T.S. Eliot*

Introduction

I N KEEPING with the basic theme of the book, this chapter will introduce the concept of *intellectual property*, particularly those aspects of it (patent, copyright, trade secret) that are useful in protecting the results of creativity and research. It is impossible to cover all aspects of intellectual property in one chapter, so the focus will be on aspects of intellectual property that are particularly applicable to researchers in their business and academic affairs.

This chapter will introduce tangible (physical) properties—that is, the inventions, works of authorship, and proprietary information (including documents and material things) that are produced in the course of research—and the intangible (intellectual) properties (particularly patents, copyrights, trade secrets) that provide a holder with a legally protectable interest in those inventions, works of authorship, and/or proprietary pieces of information. Furthermore, it will introduce the concepts of joint inventorship (in the case of inventions) and joint authorship (in the case of works of authorship). It will also introduce the importance of record-keeping as a tool to establish and document the creation of intellectual properties, and to enable researchers to deal with government agencies and/or the courts with the candor and fair exchange required to establish and maintain those properties. Still further, this chapter will explore some aspects of intellectual property (i.e., the patentability of biotechnology-based and computer-soft-ware- based inventions, and the idea of *fair use* under the copyright law) that are the subject of intense debate today. And finally, it will alert the reader to the seriousness with which the law views willful misappropriation of intellectual property. Note: In general, the intellectual properties dis-cussed in this chapter will be those protected under United States law.

While intellectual property protection does provide its holder with a legally defendable interest in the property, it also provides the public with a benefit. For example, a patent for an invention provides its holder with the right to prevent others from making, using, or selling the invention for a limited period of time (seventeen years from the patent-issue date for patents filed before June 8, 1995; twenty years from the filing date of the patent for patents filed on or after June 8, 1995), but when the patent expires, the invention passes into the public domain and is available freely for use by the public (Title 35, United States Code 154). A copyright for a work of authorship such as a writing, drawing, sculpture, or computer program provides its holder with certain exclusive rights—for example, the rights to reproduce, publish, create derivatives of, publicly display, and/or perform the work of authorship—but at the end of the copyright term (in the case of an individual author, the author's life plus fifty years), the work again enters the public domain. Moreover, even during the copyright term the public has the right to certain uses that are considered *fair* uses. Even a trade secret, which might never pass into the public domain, may provide the public with a benefit if it enables (or at least encourages) the holder to put something on the market that is valued by the public. For example, the Coca Cola formula is a closely guarded trade secret, but the public receives a benefit in terms of a popular product produced according to the formula.

Properties: Tangible and Intangible

During the research or creative process, it is common that inventions are devised; works of authorship are created; proprietary information is gener-ated; and documents and material things are produced to reflect, evidence, and record the results of the research or creative process. For example, machines and devices are conceptualized, drawn and described, and then built; research is conducted, notes and outlines are constructed, and scho-larly articles are written; a computer program is conceptualized and then implemented in a tangible form on a magnetic disc or other recording medium. Those machines, devices, notes, outlines, articles, magnetic discs, and other documents and material things can be handled, touched, and reproduced. They are the *tangible* (physical) properties that are the result or evidence of the researcher's efforts.

The *intangible* (intellectual) properties are the legally protectable interests in the inventions, works of authorship, and trade secrets that reside in or are evidenced by the physical properties. Such intellectual properties come into existence in different ways. For example, a patent for an invention is granted by the U.S. Patent and Trademark Office only after rigorous examination of a patent application filed by the inventor. On the other hand, copyright for a work of authorship comes into existence when the work of authorship is *fixed* in a tangible medium of expression, and there is a relatively simple, straightforward procedure for registering the copyright with the Library of Congress to enable the copyright holder to go to court to enforce the copyright against an infringer. With proprietary information, if the proprietor correctly protects that information as a *trade secret*, the laws of most states provide the proprietor with a right to go to court to redress a misappropriation of the trade secret (Milgrim 1.10).

Federally Protected Intellectual Property—Patents and Copyrights

The United States Constitution, Article I, section 8, provides Congress with the power "to promote the progress of science and the useful arts, by securing for a limited time to authors and inventors the exclusive right to their respective writings and discoveries." Pursuant to this authorization, Congress has enacted the federal patent law (Title 35, United States Code 1 *et seq.*) and the federal copyright law (Title 17, United States Code 101 *et seq.*).

Trade-Secret Protection

Currently, there is no federal law relating specifically to trade secrets. Under the Tenth Amendment to the Constitution, all powers not specifically granted to Congress are reserved to the states and their citizens. Therefore, protection for trade secrets is largely on a state-by-state basis.

However, most states have a generally similar notion of what a trade secret is. For example, the state of Florida defines a trade secret as a:

> formula, pattern, device, combination of devices, or compilation of information which is for use, or is used, in the operation of a business and which provides the business an advantage, or an opportunity to obtain an advantage . . . [and] . . . includes any scientific, technical, or commercial information, including any design, process, procedure, list of suppliers, list of customers, business code, or improvement thereof . . . when the owner thereof takes measures to [keep it secret] (Florida Statute 812.081).

Most states treat trade secrets as property, and protect against misappropriation of trade secrets by (1) regarding the misappropriation as a breach of a confidential relationship, as a breach of a trust, or as a form of theft; (2) allowing the proprietor to seek monetary damages and other forms of relief for the misappropriation; and (3) in many cases, punishing an intentional wrongdoer with criminal sanctions. Also, if the trade secret is exported to a foreign country or is transported across state lines, certain federal laws may be violated (Milgrim 1.10).

Types of Patents

Under the federal patent law, the most common type of patent is the *utility* patent. More specifically, the federal patent law provides that "Whoever invents or discovers any new and useful process, machine, manufacture, or composition of matter, or any new and useful improvement thereof, may obtain a [utility] patent therefor, subject to the conditions and requirements of this title" (Title 35, United States Code 101).

The foregoing section of the federal patent law also defines what is often referred to as *statutory subject matter*. If an invention is considered statutory subject matter, an inventor has a right to seek a patent for the invention. However, as will be seen in later sections, in two patent areas that are the subject of intense debate today (i.e., biotechnology-based and computer-software-based inventions) a core issue with which inventors, the U.S. Patent and Trademark Office, and the courts are wrestling is whether the subject matter for which patent protection is being sought does in fact constitute statutory subject matter.

Besides utility patents, the federal patent law also provides for both *design patents* and *plant patents*. Specifically, the federal patent law allows a designer to apply for a design patent for "any new, original and ornamental design for an article of manufacture . . ." (Title 35, United States Code 171). Moreover, the same law provides that a plant patent may be applied for by anyone who "invents or discovers and asexually reproduces any distinct and new variety of plant . . ." (Title 35, United States Code 161).

The Patent Process

The patent process starts with the filing of a *patent application*. The patent application must include a written specification (and accompanying drawings, where appropriate) that describes the invention for which patent protection is being sought in a manner that enables someone skilled in the relevant art (i.e., the relevant technology) to practice the invention. Moreover, the specification must describe the *best mode* contemplated by the inventor for practicing the invention, and must conclude with one or more *claims* that define the metes and bounds of the invention for which patent protection is being sought. Thus, the application must have *written-description* support for the invention, which demonstrates that the invention was actually in the possession of the inventor. The patent application must be *enabling* in the sense that it must be complete enough to enable someone skilled in the relevant art to practice the invention without undue experimentation, and it must disclose the best mode of the inventor (thus, the inventor cannot file for a patent on an invention but withhold from the application the inventor's best way of practicing the invention). Also, the patent application must have claims that define the invention with sufficient particularity such that they "clearly distinguish what is claimed from what went before in the [relevant] art and clearly circumscribe what is foreclosed from future enterprise" (United Carbon Company et al. vs. Binney & Smith Company).

Once an application is filed, the U.S. Patent and Trademark Office examines it for compliance with the federal patent law to see if it merits the award of a patent. During the examination, an applicant may have the opportunity to amend the application in light of any issues developed during the examination process. In essence, the examination process becomes an exchange of communications between the applicant and the Patent and Trademark Office, to arrive at a point at which (1) the Patent and Trademark Office grants the patent, (2) the applicant gives up and abandons the application, or (3) an impasse is reached, and the issue ends up in the Court of Appeals for the Federal Circuit (the federal patent court).

If during the examination process the Patent and Trademark Office discovers another patent application (or even an issued patent) from a different inventor claiming the same patentable invention as an applicant, the Patent and Trademark Office can set up an *interference*—a priority contest to determine who is entitled to the patentable invention. As will be seen later, good record-keeping, documentation, and corroboration of an inventor's work by someone other than the inventor are important in an interference.

Patent Protections

A valid U.S. patent protects the holder against unauthorized manufacture, use, or sale of a patented invention, and in many instances against importation into the U.S. of products made according to a patented invention (the *patented* invention means the invention defined by at least one of the claims of the patent) (Title 35, United States Code 271). Contrary to widely-held beliefs, a patent does *not* give its holder the right to practice the patented invention irrespective of anyone else's patent. Indeed, the practice of the patented invention of one patent could be an infringement of a broader, dominating patent of another. Thus, while a patent gives its holder the right to *exclude others* from practicing the patented invention, it does not give the holder the right to practice the patented invention free of the rights of others.

Publication and Patent Rights

Because there is normally a desire or even a requirement for researchers to publish, it is useful to comment on the legal effect on U.S. and foreign patent rights of early publication of an invention. In the United States, if an invention is described in a printed publication anywhere in the world, that publication will bar a U.S. patent application from being validly filed for the invention after one year from the date of such publication. The publication becomes what is commonly referred to as a statutory bar. However, in this case an inventor does at least have a one-year grace period from the publication date before the publication bars the application.

Things are *not* the same in most industrialized foreign countries, however, which have what is referred to as an *absolute novelty standard*. This means that an early publication *before* a foreign patent application is filed is interpreted as prior art against the foreign application. However, under an international treaty known as the Paris Convention, a U.S. applicant can get the benefit of the filing date of a U.S. patent application for foreign patent applications filed within one year of the filing date of the U.S. patent application. [Note: The Paris Convention of March 20, 1883 resulted in the foundation of the International Union for the Protection of Industrial Property, which came into force July 7, 1884.]

Thus, an early publication of an invention, *before* a U.S. patent application is filed, will become prior art against foreign applications in most industrialized countries, but it will not become a statutory bar against the U.S. patent application unless the early publication date is more than one year prior to the filing date of the U.S. patent application.

Copyright

Under the federal copyright law:

> Copyright protection subsists . . . in original works of authorship fixed in any tangible medium of expression, now known or later developed, from which they can be perceived,

reproduced, or otherwise communicated, either directly or with the aid of a machine or device. Works of authorship include the following categories:

(1) literary works [computer programs generally fall into this category];
(2) musical works, including any accompanying words;
(3) dramatic works, including any accompanying music;
(4) pantomimes and choreographic works;
(5) pictorial, graphic, and sculptural works;
(6) motion pictures and other audiovisual works;
(7) sound recordings; and
(8) architectural works

In no case does copyright protection for an original work of authorship extend to any idea, procedure, process, system, method of operation, concept, principle, or discovery, regardless of the form in which it is described, explained, illustrated, or embodied in such work (Title 17, United States Code 102).

The federal copyright law also provides that *compilations* and *derivative* works can be proper subject for copyright, but cautions that:

The copyright in a compilation or derivative work extends only to the material contributed by the author of such work, as distinguished from the preexisting material employed in the work, and does not imply any exclusive right in the preexisting material. The copyright in such work is independent of, and does not affect or enlarge the scope, duration, ownership, or subsistence of, any copyright protection in the preexisting material (Title 17, United States Code 103).

Notice that copyrightable subject matter must be *original* (i.e., a creation of the author) and must be *fixed in a tangible medium of expression*. Notice further that copyright extends only to the original material contributed by the author; it does not extend to any preexisting material employed in the work of authorship. Further, notice that copyright protection does not extend to any process or method of operation embodied in a work of authorship. Thus, if a work of authorship (e.g., a sculpture) is embodied in a useful article (e.g., a belt buckle), the copyrightable work of authorship must be able to exist (at least conceptually) apart from the useful article to be proper subject matter for copyright (Kieselstein-Cord vs. Accessories by Pearl, Inc.).

In identifying the holder of a copyright for a work of authorship, it is always necessary to identify the *author* of the work. If a work of authorship is created by an individual who is not employed by someone else, and who has not been specifically hired to create the work of authorship, the individual is regarded as the author, and the copyright belongs to the individual. However, if the work of authorship is considered a *work made for hire*, as that term is defined in the federal copyright law, the employer or person commissioning the work is considered the *author*. Though the words *employer* and *person commissioning the work* may sound straightforward, when one attempts to apply them to a particular type of employment or special business relationship they are often not so clear. Thus, the issue of whether a work is in fact a "work made for hire" must be evaluated on a case-by-case basis.

Also, as will be seen in a later section, whether a work of authorship is a *joint* work (i.e., jointly authored) can sometimes be a difficult issue to resolve.

Copyright Registration

The copyright registration process is, in general, far simpler than the patenting process. What is required is a relatively short application in which one must, among other things, identify the *author*, state whether the work was a *work made for hire*, identify any *preexisting material* incorporated in the work of authorship, and specify the author's original contribution to the work of authorship. As one might well surmise in light of the foregoing discussion, those issues may require careful attention when an application for copyright registration is prepared.

Rights Conferred by Copyright

The federal copyright law states that:

Subject to [certain exceptions] the owner of copyright . . . has the exclusive rights to do and to authorize any of the following:

(1) to reproduce the copyrighted work in copies or phonorecords;
(2) to prepare derivative works based on the copyrighted work;
(3) to distribute copies or phonorecords of the copyrighted work to the public by sale or other transfer of ownership, or by rental, lease, or lending;
(4) in the case of literary, musical, dramatic, or choreographic works, pantomimes, and motion pictures and other audiovisual works, to perform the copyrighted work publicly; and
(5) in the case of literary, musical, dramatic, or choreographic works, pantomimes, and pictorial, graphic, or sculptural works, including the individual images of a motion picture or other audiovisual work, to display the copyrighted work publicly (Title 17, United States Code 106).

A key *exception*, which will be discussed in more detail later, is the *fair use* exception. Specifically, the federal copyright law provides that:

The fair use of a copyrighted work, including such use by reproduction in copies or phonorecords or by any other means specified by that section, for purposes such as criticism, comment, news reporting, teaching (including multiple copies for classroom use), scholarship, or research, is not an infringement of copyright. In determining whether the use made of a work in any particular case is fair use, the factors to be considered shall include:

(1) the purpose and character of the use, including whether such use is of a commercial nature or is for nonprofit educational purposes;

(2) the nature of the copyrighted work;
(3) the amount and substantiality of the portion used in relation to the copyrighted work as a whole; and
(4) the effect of the use on the potential market for or value of the copyrighted work (Title 17, United States Code 107).

Notice that, unlike patent protection, which provides its holder with the right to *prevent* others from doing certain things, the federal copyright law grants to a copyright holder the *exclusive* right to *do and to authorize* others to do certain things.

Also, notice that there is a specific exception for the *fair use* of a copyrighted work, and that exception is plainly written to protect the news reporting, teaching, and research communities. Moreover, the fair use exception is obviously written to enable a court to use its sense of justice and fairness in those cases that come before it. However, as we will see in a later section, the type of fact finding, analysis, and balancing of equities a court may have to go through to determine if the fair use exception applies can be difficult.

Finally, before leaving the subject of the rights conferred by copyright, it is important to comment briefly on an *author's moral rights*. In many foreign countries, an author has the rights of recognition and attribution, can prevent the distortion or distribution of the work of authorship if the work no longer represents the author's views, and can prevent the use of the work or the author's name in a way that affects the author's professional standing. Some similar rights are provided under the federal copyright law to authors of visual works (Title 17, United States Code 106A).

The Importance of Properly Identifying Inventors and Authors

Why is it so important to identify the inventors of a patentable invention or the authors of a work of authorship? One reason is that the true inventors must be identified in a patent application, and the authors must be identified in an application for registration of copyright in a work of authorship (Title 35, United States Code 102f). Another reason is that unless the inventors or authors have conveyed or agreed to convey (transfer) their rights to someone else (e.g., an employer), the inventors or the authors, as the case may be, may have an ownership interest in the patentable invention or the copyright in the work of authorship.

In addition, in a patent application, not only must each inventor be identified, but each inventor must also sign an oath (or declaration) stating, under penalty of fine or imprisonment, that that person is in fact an inventor of the subject matter for which patent protection is being sought (as discussed earlier, this means the person must be an inventor of at least one of the claims of the patent application).

Still further, in an application to register a copyright in a work of authorship, it is necessary to identify each person who is an author, and to state whether that person is an author under the "work made for hire" doctrine (discussed earlier). It is also necessary to describe the contribution to the work of authorship made by each author.

These are but a few of the reasons why it is important to properly identify inventors and authors; there are many more.

Joint Inventorship and Joint Authorship

The federal patent law does not have a definition of joint inventorship. Rather, it provides the following statement:

> Where an invention is made by two or more persons jointly, they shall apply for patent jointly and each make the required oath [declaration] . . . . Inventors may apply for a patent jointly even though (1) they did not physically work together or at the same time, (2) each did not make the same type or amount of contribution, or (3) each did not make a contribution to the subject matter of every claim of the patent (Title 35, United States Code 116).

Plainly, when two or more persons are collaborating on all phases of a research project, and jointly conceptualize and build an invention, they are joint inventors. More often, things are not quite so straightforward, so that careful attention must be paid to exactly how and by whom an invention was conceived and then implemented, in order to determine who is (or are) the true inventor(s).

The federal copyright law does provide a definition of *joint work*. It states that a joint work is "a work prepared by two or more authors with intention that their contributions be merged into inseparable or interdependent parts of a unitary whole" (Title 17, United States Code 101). Notice that to be a joint work the contributions of the authors can be *either* inseparable *or* interdependent. Moreover, the authors do not have to work together, or even know each other, so long as each *intends* his or her contribution to be merged into inseparable or interdependent parts of a unitary whole.

From the previous discussion it is plain that inventorship and authorship are *not* the same thing. If two or more persons collaborate to create an invention, but only one of them takes the laboring oar to create a journal article about their research, the one who wrote the journal article could well be the author of the article, even though the invention is joint.

The Importance of Record-Keeping

Inventors typically record and document their research efforts. In the creation of books and articles, authors typically do their research, decide what they need to attribute to other sources, and use their creative resources to organize and prepare their written works.

With inventions, documentation is important to establish that the person(s) claiming to be inventor(s) is (are) in fact inventor(s), and to establish how and when the invention(s) came into existence. In an interference, where two different applicants approach the U.S. Patent and Trademark Office claiming the same patentable invention, record- keeping and corroboration as to when each applicant *first* (1) conceived the patentable invention, (2) disclosed it to someone else, (3) prepared a written description, (4) prepared drawings, and (5) built and successfully tested the invention, may be critically important to winning the interference.

In the case of a work of authorship, proper documentation may be important for identifying the true authors, and may be useful in defending an author against a charge of infringement. For example, it is at least theoretically possible that, if two different authors working independently in different locales—neither having access to the other's work—create two works that are so similar that it would be natural to assume that one was copied from the other, it may be critical for the author who is charged with copying to be able to establish, by appropriate records and documentation, the manner and time frame in which that author created the work.

Candor and Fair Dealing in Connection with Intellectual Property

In interacting with government agencies or the courts in connection with intellectual property, candor and fair dealing are important. A duty of candor and fair dealing is specifically imposed upon inventors and their attorneys in their approach to the U.S. Patent and Trademark Office (37 CFR, 1.56). This means that inventors and their attorneys must put before the Patent and Trademark Office everything they know of that would be material to their patent applications. Thus, if an inventor is aware of material prior art against which an invention should be properly evaluated, that prior art should be put before the U.S. Patent and Trademark Office. If an inventor has some data to support the proposition that an invention is useful for a particular purpose, but also has data suggesting the opposite, the duty of candor and fair dealing generally requires that all of that data be fairly presented to the Patent and Trademark Office, so that the Office can make an informed decision on the patentability of the invention (Cosden Oil & Chemical Company vs. American Hoechst Corporation). The duty of candor and fair dealing also requires that the inventor(s) be properly identified. Indeed, as indicated earlier, there can be severe consequences for a person falsely claiming to be an inventor when the person knows he/she is not an inventor.

Though the phrase *candor and fair dealing* is not specifically used in the federal copyright law, an applicant for copyright registration must prepare the application in good faith, and must certify, on the applicant's knowledge and belief, that the representations in the application are true. Courts can hold a copyright registration invalid and/or unenforceable where fraudulent representations are made in the copyright registration (Russ Berrie & Company, Inc. vs. Jerry Elnser Co., Inc.). Therefore, candor and fair dealing are plainly necessary in the preparation of an application for copyright registration.

Of course, candor and fair dealing are also at the essence of a trade secret. If a person is entrusted with access to what the person knows to be a trade secret, that person is put in a position of trust in protecting the trade secret, and a breach of that trust is what normally leads to misappropriation of the trade secret.

Fair Use

In an earlier section, it was noted that *fair use* of a work of authorship is not an infringement of copyright in the work of authorship. Moreover, as noted, the key factors to be considered in determining whether a use is a "fair use" are:

(1) the purpose and character of the use, including whether such use is of a commercial nature or is for nonprofit educational purposes;
(2) the nature of the copyrighted work;
(3) the amount and substantiality of the portion used in relation to the copyrighted work as a whole; and
(4) the effect of the use on the potential market for or value of the copyrighted work.

On face value, those factors appear to be simple, straightforward, and equitable. However, applying them to a particular situation is not always simple or straightforward. The difficulties and the delicate balancing act that a court must sometimes engage in while weighing the fair-use factors are reflected in a case commonly known as the *Texaco case* (American Geophysical Union vs. Texaco Inc.).

In the Texaco case, Texaco Inc. was accused of copyright infringement because of the copying, by or for its researchers, of articles from technical publications to which Texaco subscribed. Texaco would circulate technical journals to its researchers, who would in turn either copy the articles in which they were interested or have Texaco's library make the copies. In the one researcher's files that were used by the court (with the agreement of all parties) to crystalize the issue, were to be found copies of eight articles from the *Journal of Catalysis*. The evidence indicated to the court that the researcher did not use most of the copies of the articles when he prepared them; instead he kept them in his files for future reference. Texaco argued, among other things, that the copying was within the fair use exception, because the purpose of copying the articles was to transform them into a form more suitable for supporting the research of its scientists. The court, after an exhaustive analysis of the four factors, and over a strong dissent by one of its members, found that Texaco's actions were *not* within the fair use exception. What should be of particular interest to researchers is how the court addressed the first factor, that is, the purpose and character of the researcher's copying of the articles. The court found that, contrary to Texaco's arguments, the researcher had copied the articles primarily to create his own personal files or archives, and had not used most of the copied articles, so that in that sense the copying did not transform the articles into a form more suitable for scientific research.

In light of the *Texaco case*, the copying of journal articles principally for the purpose of creating personal archives is risky.

Biotechnology-Based and Computer-Software-Based Inventions

Considerable research is being conducted in the areas of biotechnology and computer software. Coincidentally, those are also areas generating important issues within the U.S. Patent and Trademark Office and in the courts. In biotechnology, one particularly important issue is whether a claimed invention is "useful". Another is whether a patent application is "enabling" for a claimed invention. In an earlier section it was noted that in order to constitute *statutory subject matter* an invention must be *useful*, and that a patent application must *enable* one skilled in the relevant art to practice the claimed invention.

In the area of utility, the U.S. Supreme Court in 1966 held that a claim for producing a product for which the applicant could not prove a specific utility was not based on a useful invention, stating, among other things "until the process has [produced] a product shown to be useful, the metes and bounds of the [patent] monopoly are not capable of precise delineation" (Brenner vs. Manson).

Clearly, that type of rationale creates significant challenges to researchers in the biotechnology area in terms of demonstrating utility for their inventions. It also demonstrates the challenges confronting biotechnology researchers and their patent counsel in satisfying the enablement requirement of the federal patent law. These are issues that are currently being wrestled with by the U.S. Patent and Trademark Office and the courts, and which are likely to continue to challenge inventors and their patent counsel as biotechnology-based inventions move into even newer, as yet unexplored, areas.

In *computer-software-based inventions*, the courts and the U.S. Patent and Trademark Office are also wrestling with the issue of whether a claimed invention is directed to statutory subject matter, but the factual and legal inquiries underlying that issue come from a different direction. Specifically, in the first of a series of landmark cases, commonly referred to as the *Benson case*, the U.S. Supreme Court determined that a patent claim for a method of converting binary-coded decimal numbers into binary form according to the inventor's mathematical formula, which was structured to cover the method in a general-purpose digital computer, was tantamount to a claim on the mathematical formula (or algorithm) itself, and denied patent protection on the ground that the claim did not define statutory subject matter within the meaning of the federal patent law (Gottschalk vs. Benson).

In the last of that series of cases, commonly referred to as the *Diehr case*, the U.S. Supreme Court, by a narrow majority, found that a claim for a rubber-curing process, in which a computer was programmed to receive data about temperature and timing and to use the data in a formula to determine the time to open a curing press, *did* define statutory subject matter even though the general process for curing rubber was not new and the mathematical formula itself was not new either. The court reasoned that the claim was not for the mathematical formula; it was for the use of that formula in connection with a process for curing rubber (Diamond vs. Diehr).

So, from *Benson* we know that if a patent claim is determined to effectively cover the solution to a mathematical formula (an algorithm) it is not statutory subject matter. On the other hand, from *Diehr* we know that if a claim is directed to a machine, process, etc. and happens to also recite the use of a mathematical formula, so long as the effect of the claim is not to cover just the solution to the formula, the claim is directed to statutory subject matter. Though this may seem straightforward, it is certainly not.

Recently the Court of Appeals for the Federal Circuit (the federal patent court) entered fully into this issue in a landmark case commonly referred to as *Alappat*. In *Alappat*, the court in a six-to-five decision found statutory subject matter in a claim directed to what the majority characterized as a machine, namely, a rasterizer "for converting vector list data representing sample magnitudes of an input wave form into anti-aliased pixel illumination intensity data to be displayed on a display means . . ." whereas the dissenters, in disagreeing that the claimed invention was directed to statutory subject matter, found that:

> The "rasterizer" as claimed is an arrangement of circuitry elements for converting data into other data according to a particular mathematical operation . . . . The claimed "rasterizer" ends with other specific "data"—an array of numbers . . . . The end data of the "rasterizer" are a predetermined and claimed mathematic function of the two input numbers (In re Alappat).

A divided court, two groups of distinguished judges, the same application, and two different ways of looking at things—clearly, inventors and their patent counsel have much to do regarding computer-software-based inventions.

Piracy or Misappropriation

Piracy or misappropriation of other people's inventions, works, or trade secrets is serious business:

. . . so serious that the federal copyright law has criminal penalties for intentional or willful violation of a copyright owner's rights (Title 17, United States Code 506);

. . . so serious that intentional misappropriation of a trade secret can give rise to criminal sanctions (Milgrim 1.10);

. . . so serious that the Software Publisher's Association has been labeled "the software police" because of its aggressiveness in pursuing those who copy the software of its members;

. . . so serious that the federal patent law provides a court with the power to triple the damage awards against willful infringers of patents (Title 35, United States Code 284).

Thus, while intellectual properties exist to reward the creative endeavors of inventors and authors, it is equally important that those who are exposed to the intellectual properties of others recognize that any misappropriation of those properties, before they enter the public domain, can have serious consequences.

CASE STUDY 8 A

Al and Go and the Algorithm

Al and Go jointly develop an algorithm describing the way light is transmitted by a new type of glass substance. They decide it would be a good idea to create a computer program to solve the algorithm, and to use the algorithm to create a new mirror for a telescope.

Al creates a computer program to solve the algorithm. Go creates a new mirror for a telescope using the algorithm, and shows the mirror to Al. Al decides to write a journal article on all of their work, writes the article, and asks Go to review and edit the article. Go does that, but also inserts his name as a co-author. Al takes Go's name off the article, but suggests to Go that they file a patent application on the new mirror, listing both of them as joint inventors.

Go, on learning Al has taken Go's name off the article, objects to Al's being included as a joint inventor on the patent application. Al then decides to file a patent application on the use of Al's computer program to determine the profile of Go's telescope mirror.

Questions

1. Who is the author of the article? Al? Al and Go? (Hint: who took the laboring oar in the writing of the article?)
2. Who invented the mirror? (Hint: Was Al's computer program required to create the mirror?)
3. Who invented the computer program?
4. Who can file a patent application? Al? If so, on what? (Is a computer program that solves an algorithm patentable? Is the use of that program to create Go's telescope mirror patentable?) Can Go file a patent application? If so, on what? Can both file a patent application? If so, on what?

CASE STUDY 8 B

The Gene-Nome Project

Gene and Nome are biochemical researchers at ABC, Inc. While working together on a project to identify a cure for the HIV virus, they isolate a compound they suspect might be an immunosuppressant for HIV. Gene also notices something about the compound that is very similar to compounds used to reduce arthritis inflammation, and begins to do some literature research on possible use of the compound (which they name GTO) as an anti-inflammatory agent for arthritis.

Gene goes to ABC's library, makes copies of journal articles on anti-inflammatory agents for arthritis and, based on the journal articles, develops research procedures for evaluating GTO as an anti-inflammatory agent for arthritis. Gene records the procedures in a notebook and shows the articles to Nome, who agrees that GTO has the potential to be an anti-inflammatory agent for arthritis. Gene suggests filing an invention disclosure on the use of GTO as an anti-inflammatory agent for arthritis, naming Gene as the inventor and Nome as a corroborating witness. Nome expresses some surprise because Gene and Nome together discovered the GTO compound originally. They go to their supervisor at ABC to try to resolve the issue, but the supervisor tells them to get back to their work on HIV research and take up this new project only after they have finished the first one. Gene puts aside his collection of journal articles and keeps them in a file which Gene intends to refer to when the new project resumes.

Questions

1. Has Gene, or have Gene and Nome together, discovered a useful compound for which patent protection can be sought? (Hint: Have either of them demonstrated the usefulness of GTO for anything?)
2. If they have, who is the inventor? And of what?
3. If Gene is the sole inventor of something, can Nome help Gene corroborate Gene's invention? (Even if you decide Gene has not yet invented anything, it is still worthwhile to pursue this question assuming Gene *has* invented something, and to think about the issue of corroboration for another's invention.)
4. Was Gene entitled to copy the articles from journals in ABC's library without infringing the journal publisher's copyrights? (Hint: Remember the Texaco case.)
5. If Gene was entitled to copy the journal articles, is Gene entitled to keep them as a collection?

This case study is intended to raise the issue of the patentability of compounds for which practical utility has not been demonstrated. It is also intended to introduce a variation on the Texaco case, in that (1) when Gene copied the articles, it was with the specific intent to develop procedures to evaluate GTO, (2) Gene did develop those procedures, and (3) only when they were removed from the new project did Gene put the articles aside and retain them as a collection.

Recommended Reading

For emphasis on biotechnology-based inventions, see:

A.P. Gershman, *The Ethics of Biotechnology: The Kosher Goldfish.* The Law Works 2:3, Mar. 1995, pp. 13–19.

A.F. Konski, *The Utility Rejection on Biotechnology and Pharmaceutical Prosecution Practice.* Journal of the Patent Office Society, Nov. 1994, pp. 821–831.

C.A. Michaels, *Biotechnology and the Requirement for Utility in Patent Law.* Journal of The Patent Office Society, Apr. 1994, pp. 247–260.

W.D. Woessner, *Patenting Life—Transgenic Animals.* The Law Works 1:12, Dec. 1994, pp. 8–10.

For emphasis on trade secrets, see:

R.M. Milgrim, *Milgrim: The Law of Trade Secrets.* Matthew Bender, New York 1987. (See particularly Chapter 1 concerning the notion of trade secrets as property, and a general discussion of how the law protects trade secrets.)

For emphasis on copyrights, see:

B.M. Nimmer, D. Nimmer, *Nimmer on Copyright.* Matthew Bender, New York 1994. (See particularly Chapters 1–6, 8.)

For emphasis on computer-software-based technology, see:

R.T. Nimmer, *The Law of Computer Technology.* Warren, Boston 1985.

J.L. Rogitz, *'Scoping' the Alappat Decision!* The Law Works 1:10, Oct. 1994, pp. 12–16.

J.L. Rogitz, *And You Thought You Understood Alappat.* The Law Works 2:2, Feb. 1995, p. 19.

Works Cited

In re Alappat, 33 F, 3d 1526 (CAFC 1994).

American Geophysical Union vs. Texaco Inc., 37 F, 3rd 881 (2nd Cir. 1994).

Brenner vs. Manson, 383 US 519, 16 L. Ed. 2d 69, 86 S. Ct. 1033 (1966).

37 CFR 1.56 (Code of Federal Regulations).

Cosden Oil & Chemical Company vs. American Hoechst Corporation, 214 U.S.P.Q. 244 (D.Del. 1982).

Diamond vs. Diehr, II, 450 US 175, 67 L. Ed. 2d 155, 101 S. Ct. 1048 (1981).

Gottschalk vs. Benson, 409 US 63, 34 L. Ed. 2d 273, 93 S. Ct. 253 (1972).

Kieselstein–Cord vs. Accessories by Pearl, Inc., 632 F.2d 989 (2d Cir. 1980).

R.M. Milgrim, *Milgrim: The Law of Trade Secrets*. Matthew Bender, New York 1987.

Russ Berrie & Company, Inc. vs. Jerry Elnser Co., Inc., 482 Fed. Supp. 980 (SDNY 1980).

Title 17, United States Code—Copyrights 101 *et seq.*

Title 35, United States Code—Patents 1 *et seq.*

United Carbon Company et al. vs. Binney & Smith Company, 317 U.S. 228 (1942).

CHAPTER NINE

Misconduct in Scientific Research

Bryan S. Burgess

> The object of scientific knowledge exists necessarily; and it is therefore eternal, since all things that, in the strict sense, exist necessarily are eternal; and eternal things are not subject to creation and destruction.
>
> *Aristotle, Ethics Book VI*

Introduction

ARISTOTLE HELD that science is one means through which the soul comes to the truth, defining it as a disposition that has to do with demonstrative knowledge. Scientists, then, are seekers and devotees of truth—using systematic methods to observe and experiment with facts and the relationships among those facts, and to develop theories to order and integrate the facts, pursuant to their search for a systematic organization of knowledge about the universe. The essential, distinguishing characteristic of science, in contrast to other branches of knowledge, is that scientific ideas can and must be proven; they must be tested and verified as true. The topic of this chapter, misconduct in scientific research, is concerned with aberrant behaviors and actions of scientists that detract from the integrity of this quest for eternal truths, and with governmental and institutional guidelines and procedures that have evolved to identify, review, and punish such misconduct when it occurs.

For centuries, the scientific community has upheld and perpetuated traditions and standards in the research process founded on generally accepted ethical principles such as honesty, integrity, objectivity, and collegiality. This shared set of values has fostered such normative rules for responsible research as respect for primary data, adherence to verifiable research methods, and the reporting of negative findings. Consequently, the community of scientists has historically enjoyed a high degree of autonomy and self-governance. Scientists have been able to count on each other and on the controls and safeguards inherent in the process of performing, evaluating, and reporting scientific research to ensure integrity in the research enterprise. Unreliable work has been identified through various self-regulating and self-correcting mechanisms of the scientific process, such as efforts of other scientists to replicate work and peer review of data and publications. The ethics and practices of this honor system were not legislated or promulgated formally, but were taught and impressed through succeeding generations of scientists (National Academy, p. 1).

However, over the past two decades, the capability of the community of scientists to police itself and protect the integrity of the research process has been subject to doubt and attack as a consequence of several developments. The rapid and significant growth in the the U.S. research system and increasing emphasis on the commercial application and exploitation of research products has resulted in new stresses. The greater size and specialization of research teams has often strained the ability of individual scientists to maintain the traditional high level of personal involvement and familiarity with the work of their colleagues and subordinates. Pressures to achieve and survive in academia have produced disputes and schisms relative to the allocation of credit, recognition of intellectual property rights, competition for funding, and so forth. At the same time, the modern laboratory has been brought under increasing scrutiny and criticism by investigative reporters, politicians, and disgruntled taxpayers. Public reports in the media have brought to the attention of the scientific community, the public at large, and the federal government a disturbing number of instances of scientists who reported measurements they never made, altered research results, or plagiarized the work of others. These circumstances have all led to changing social and governmental expectations concerning the accountability of scientists and their institutions for research endeavors.

The scientific misconduct issue was brought into the national limelight in 1974 with the "painted mouse" incident involving William Summerlin, a physician–researcher at the Sloan–Kettering Institute for Cancer Research in New York. Dr. Summerlin used a felt-tipped marker to paint a black patch on white mice to simulate successful skin grafts from black to white mice. In the early 1980s there were several additional widely publicized cases involving allegations of misconduct against researchers at major U.S. institutions, such as fabrication of professional credentials, falsification of patient records and raw data, and plagiarism.

These cases produced alarm that the incidence of misconduct was increasing, and serious concern that there were no established formal mechanisms to handle allegations of misconduct in science and maintain integrity in the research environment. This in turn precipitated a series of Congressional hearings beginning in 1981. At the early hearings, the scientific community's view that the problem was internally manageable prevailed. However, concerns about misconduct persisted as more cases and reports of poor handling of allegations were made public. This led to renewed Congressional attention to the subject, focusing on the roles of the Public Health Service (PHS)/National Institutes of Health (NIH) and the National Science Foundation (NSF)—assuring that allegations of misconduct would be reviewed and reported properly and that the integrity of federally funded research would be maintained.

In 1985 Congress passed a bill over the veto of President Ronald Reagan that formally directed PHS and NSF to publish regulations requiring grantee institutions to have systems in place for handling allegations of misconduct, and to report the outcomes of these investigations to the agencies in a timely fashion. PHS took several years to implement this legislation, promulgating its final rule in August 1989 (42 CFR Part 50). NSF promulgated its final rule in July 1987 (45 CFR Part 689).

Both PHS and NSF adopted rules that assigned the primary responsibility for preventing and detecting misconduct in research, and for handling allegations when they arise, to the institutions where the research was conducted. Institutions were required to establish policies and procedures regarding misconduct in research as a condition of receiving federal funds.

While the intensive administrative process of developing regulations was ongoing at PHS, the continuing rise in publicly reported misconduct cases and concerns about the oversight of allegations produced serious questions as to whether the 1985 legislation had gone far enough. Consequently, in March 1989, the Department of Health and Human Services (HHS) established a two-level organization to manage the problem of scientific misconduct and to foster responsible conduct of research. An Office of Scientific Integrity (OSI) was created in NIH, and an Office of Scientific Integrity Review was established in the Office of the Assistant Secretary of Health (a higher organizational level in HHS) to oversee OSI. These new offices were given responsibility for assuring compliance of universities and other research institutions with the PHS regulations.

As a starting point in a discussion of misconduct in scientific research, it is useful to consider the dimensions of the problem. There is a wide variance in perceptions of how much misconduct there is in the research system. The scientific community has argued that research misconduct is a most rare occurrence, while critics have claimed that it is much more common than scientists admit. This extreme difference in views is exemplified by, on the one hand, *Science* editor Daniel E. Koshland's proclamation in 1987 in *Science* (Vol. 235) that "99.9999 percent of [scientific] reports are accurate and truthful" (as quoted in Teich and Frankel), and on the other hand, Congressman John D. Dingle's (D.-MI) assertion in his opening statement to the 1988 House oversight committee hearing on fraud in NIH grant programs that there is "growing reason to believe that we are only seeing the tip of a very unfortunately dangerous and important iceberg" (as quoted inTeich and Frankel).

Though the actual incidence of misconduct in research is unknowable, there are some data available on cases that have been brought to light. One sociologist who surveyed the issue identified twenty-six publicly reported cases of research fraud in the period 1980 to 1987 (Project on Scientific Fraud, pp. 37–86). All of these cases involved allegations of data fabrication or serious misrepresentation that were either admitted by the perpetrator or proved to the satisfaction of an investigating body. Other indications of prevalence may be obtained from the caseloads and findings of the federal offices established to monitor and investigate cases of fraud and misconduct in science. During the period from March 1989 to March 1991, more than 200 allegations of misconduct in science were recorded by U.S. government offices. From this number about thirty cases resulted in confirmed findings of misconduct.

Definitions of Misconduct

Of course, the answer to the question of how common misconduct is in research depends largely on how one defines *misconduct*, that is, what should be included in and excluded from the set of behaviors that is regarded as misconduct in scientific research.

The NSF regulations define misconduct to mean "fabrication, falsification, plagiarism, or other serious deviation from accepted practices in proposing, carrying out, or reporting results from activities funded by NSF" [45 CFR 689.1 (a)]. The HHS/PHS regulations similarly define misconduct to mean "fabrication, falsification, plagiarism, or other practices that seriously deviate from those that are commonly accepted within the scientific community for proposing, conducting, or reporting research" (42 CFR 50.102). Although these regulatory definitions encompass a problematic "grey area" regarding other deviations discussed later, they reflect the ethical canons of the scientific community that plagiarism, fabrication, and falsification of data are misconduct. What do these terms mean?

Plagiarism

Plagiarism is using and putting forth the ideas, words, or designs of another as one's own without giving appropriate credit to the originator. Black's Law Dictionary defines plagiarism as "the act of appropriating the literary composition of another, or parts or passages of his writings, or the idea or language of the same, and passing them off as the product of one's own mind" (Black's, p. 1308). This is probably the most frequently alleged type of scientific misconduct. However, the application of the definition of plagiarism in the setting of scientific research can be problematic, because there is so much use of the words, methods, and ideas of preceding researchers that the meaning of plagiarism in practice depends on the applicable standards of attribution and the degree of adherence in a particular case. Further, it may be debated whether there exists within the scientific community a common understanding as to the standards of attribution and how one distinguishes methods and ideas that have been absorbed into the public domain and one's own thinking from those that should not be used except with attribution to the originator.

Apart from these questions, though, the community of scientists as well as institutions and government officials have had no difficulty recognizing plagiarism committed in conjunction with the preparation and submission of proposal documents for research funding. The highly competitive and proprietary nature of researchers' efforts to secure financial sponsorship has engendered zero tolerance for the unauthorized appropriation and use of another's work product in this context.

As is the case with other types of misconduct, a failure to give credit to one's sources may be the result of error, such as sloppy or lazy writing or editing, or it may be an intentional effort to claim credit for the work of another. To reach a judgement on whether plagiarism has been committed requires a careful and objective examination of the facts and a conclusion about the standards of attribution and the state of mind of the alleged plagiarist.

Fabrication, Falsification, and Fraud

Fabrication means making up data or results. Black's Law Dictionary defines *fabricate* as "to invent, to devise falsely:"

> To fabricate evidence is to arrange or manufacture circumstances or *indicia*, after the fact committed, with the purpose of using them as evidence, and of deceitfully making them appear as if accidental or undesigned; to devise falsely or contrive by artifice with the intention to deceive. Such evidence may be wholly forged and artificial, or it may consist in so warping and distorting real facts as to create an erroneous impression in the minds of those who observe them and then presenting such impression as true and genuine (Black's, p. 704).

Falsification means changing or altering data or results. Black's Law Dictionary defines *falsify* as "to counterfeit or forge; to make something false; to give a false appearance to anything; to make false by mutilation or addition; to tamper with; as to falsify a record or document" (Black's, p. 726).

It should be noted that federal regulations and most institutional policy statements do not use the term *fraud*. In a precise legal sense, fraud means an intentional misrepresentation or perversion of truth for the purpose of inducing another to rely and act on it to his legal injury. This evidentiary standard (proof of intent as well as harm done by the deception) is poorly suited to methods of scientific research. In many instances of plagiarism, there is arguably no damage to the individual plagiarized unless the plagiarist publishes first, and even deliberate falsification of data arguably might not meet the legal definition of "fraud" in a situation where the faker publishes imagined experimental data to prove an otherwise established scientific result or conclusion. The concepts of fabrication and falsification involve a perversion of truth, but do not require third party reliance and injury, and thus are more apt legal standards of conduct, consistent with the imperative of absolute integrity and truth in scientific endeavors.

Examples of fabrication or falsification are reporting experiments, measurements, or statistical analyses never performed; manipulating or altering data or other manifestations of the research to achieve a desired result; falsifying or misrepresenting background information, including biographical data, citation of publications, or status of manuscripts; selective reporting, including deliberate suppression of conflicting or unwanted data without scientific or scholarly justification; and failure to perform research supported by a grant while stating in progress reports that active progress has been made.

Other Practices

Beyond the types of blatant misbehavior discussed above are other practices that deviate from ethical norms as exemplified in the practices of the scientific community. As stated previously, the PHS regulations approach this "grey area" by defining misconduct to include "other practices that seriously deviate from those that are commonly accepted within the scientific community for proposing, conducting, or reporting research" (42 CFR 50.102), while the NSF definition similarly includes "other serious deviation from accepted practices in proposing, carrying out, or reporting results from activities funded by NSF" [45 CFR 689.1 (a)].

There are different perspectives regarding the merit and meaning of this component of the regulatory definitions of misconduct, and the adequacy of the current process to achieve fair and consistent determinations in cases of this type of misconduct in scientific research.

Concern over the intrusion of external bureaucracies and abuse of flexibility has caused many scientists to object strenuously to the provisions regarding other deviations. They have argued against any nonspecific prohibition of unacceptable scientific conduct out of concern that it represents an appeal to orthodoxy, and might be used against unconventional (but not improper) science. It has been noted that the most pioneering research necessarily deviates from that commonly accepted in the scientific community, and that an overly zealous or literal interpretation and enforcement of those provisions by non-scientist bureaucrats could seriously frustrate the progress of science. On the other hand, the regulators charged with responding to the Congressional call to action knew that it was impossible to list all possible bad deeds, and that institutions needed reasonable flexibility to respond to the various kinds of egregious acts that have a direct negative impact on the integrity of the research process.

Handed this responsibility to define misconduct in their institutional policies, institutions have been forced to grapple with similar concerns. Thus, some institutional policy statements recognize a need to establish and disseminate, in this "grey area", specific standards of research conduct so that scientists at all levels will know what is expected of them, and a need to strengthen the institution's hand in imposing sanctions for violations. These policy statements make clear the boundaries between acceptable and questionable conduct by specifically enumerating and describing the various practices that are deemed to deviate from commonly accepted norms of responsible research. Other institutional policy statements mirror the federal rules, and do not provide a detailed codification of questionable or unacceptable practices. A pitfall of this approach is the perilous assumption that acceptable research practices are universally understood. For even the most obvious examples of good laboratory procedures (e.g., using ink in notebooks) a common understanding may not exist across the research community. In addition, a basic element of due process is the provision of adequate notice to the target community of the conduct that is proscribed; researchers should not have their reputations and careers placed in jeopardy by an unwitting violation of unpromulgated "standards" of research conduct.

The type of behavior that is easily placed in this category involves deviations from "commonly accepted practices" as defined by other applicable laws. Examples of this are the violation of federal, state, or institutional rules on research involving human and animal subjects, recombinant DNA, new drugs and devices, and radioactive materials.

It should be noted that certain other forms of illegal behavior are clearly not unique to the conduct of science, although they may occur in a laboratory or research environment. Such behaviors, which are subject to other applicable legal procedures and penalties, include sexual and other forms of harassment; gross negligence; vandalism, including tampering with research experiments or instrumentation; violations of conflict of interest laws; and misuse of funds or other fiscal misconduct.

Some other forms of misconduct are directly associated with misconduct in research. Among these are cover-ups of research misconduct; reprisals against whistle-blowers; malicious allegations of misconduct in research; and violations of due-process protections in handling complaints of misconduct. These other forms of misconduct are also subject to appropriate legal procedures and penalties.

Apart from these deviations from clearly established legal requirements, there remains the "common law" or community lore of other questionable practices that may be considered misconduct. Examples of such deviant practices include:

- Abuse of confidentiality with respect to unpublished materials, including use of ideas and preliminary data gained from access to privileged information through the opportunity for editorial review of manuscripts submitted to journals, and peer review of proposals being considered for funding by agency panels or such internal committees as the Institutional Review Board or Institutional Animal Care and Use Committee.
- Failing to retain significant research data for a reasonable period. On occasion, an inability to provide primary data may give rise to the inference that the data do not and never did exist.

- Maintaining inadequate research records, especially for results that are published or relied on by others.
- Conferring or requesting authorship on the basis of a special service or relationship that is not significantly related to the research reported in the paper. Many scientific journals have adopted policies regarding authorship listings.
- Other improprieties of authorship, such as improper assignment of credit (excluding others who deserve credit or including others without their knowledge), misrepresentation of the same material as original in more than one publication, submission of multi-authored publications without the concurrence of all authors, and incomplete citation of previously published work.
- Refusing to give peers reasonable access to unique research materials or data that support published papers.
- Using inappropriate statistical or other methods of measurement to enhance the significance of research findings, (e.g., skewed selection of data to hide or disguise observations that do not fit the researcher's conclusions).
- Inadequately supervising research subordinates or exploiting them (e.g., accepting responsibility for supervising unrealistic numbers of people and projects; failing to give appropriate instruction about respect for data and procedures).
- Misrepresenting speculations as fact, or releasing preliminary research results, especially in the public media, without providing sufficient data to allow peers to judge the validity of the results or to reproduce the experiments.
- Improper reporting of the status of subjects in clinical research (e.g., reporting the same subjects as controls in one study and as experimental subjects in another).
- Bias in peer review of proposals and manuscripts.

As previously noted, views and institutional policy approaches vary as to whether these questionable practices constitute serious deviations from accepted practices and thus are to be deemed "misconduct." Where institutions do not specifically prohibit these behaviors, it is left to the judgement of the reviewing officials, institutional and governmental, to determine whether the behavior in question constitutes misconduct. A judgement of this type is based on an objective analysis of all facts and circumstances in the particular case, typically including consideration of such factors as whether there is intent to deceive, whether the behavior was deliberate or merely careless, whether the behavior damages the integrity of the research process to a sufficient degree, and whether the behavior was an isolated event or part of a pattern.

Distinguishing Error From Misconduct

The PHS regulations provide that misconduct "does not include honest error or honest differences in interpretations or judgements of data" (42 CFR 50.102), and most institutional policies contain a substantially similar provision. Black's Law Dictionary defines *error* as "a mistaken judgement or incorrect belief as to the existence or effect of matters of fact, or a false or mistaken conception or application of the law" (Black's, p. 637). As observed in an article by Patricia K. Woolf in the *Journal of the American Medical Association* in 1988, the exclusion of honest error from the definition of misconduct recognizes that "Deliberate misrepresentation of scientific data is reprehensible and can never be condoned. But error is an unavoidable concomitant of vigorous research. The brash vitality that has characterized American science requires a willingness to make mistakes and to correct them when necessary. Howard Gruber's description of science is apposite: 'The power and beauty of science do not rest on infallibility which it has not, but on corrigibility without which it is nothing' " (Woolf, p. 1939).

Accordingly, the PHS regulations and institutional policies recognize that there should be no moral stigma attached to honest mistakes or disagreements. As discussed above, there is real danger that confusion between honest error and intentional misconduct may inhibit the openness and skepticism that have always characterized the scientific community. Yet, it can be very difficult to distinguish intentional falsification or other culpable error from an innocent mistake in a given case. There is a murky area between the honest errors of diligent scientists and the lies of those who claim results that they know to be false or appropriated from others. In a particular case, observers and reviewers may have differing perceptions as to where in the "grey zone" the boundary belongs. Also, it may be impossible to assess objectively whether there is intent to deceive. To conclude that an untruth is purposeful fabrication and not a mistake requires informed and critical judgement. Of course, an honest error may cross the boundary into misconduct when accepted research practices and standards for replication and confirmation are ignored. Systematic carelessness and violations of good laboratory practices may not be excused as simple honest error. Haste to publish preliminary data or the presentation of hypotheses and speculations as facts without a reasonable amount of confirmation in the lab may be viewed as culpable conduct rather than excusable error. In addition, an initially honest mistake may become misconduct if, on discovery, the researcher refuses or fails to issue a retraction or take such other corrective action as is appropriate under the circumstances (Friedman).

Governmental and Institutional Review of Misconduct

This chapter will next discuss the essential features of typical institutional policy and process, consistent with the requirements of federal regulations, for reviewing, dealing with, and reporting possible misconduct in research.

Identifying and Reporting Misconduct

Many institutional procedures provide that any person who has reason to believe that an individual has engaged in an act of research misconduct at the institution *should* report that act to the designated internal or external officials for review. The NSF has published a "Dear Colleague" letter that comments as follows on this duty to report:

> While it is sometimes unpleasant to report misconduct that you observe, it is essential to do so. Only in that way can the research community keep its own house in order and maintain both integrity and public confidence in science and engineering. It is not necessary for you to have complete evidence of the misconduct: If you have any substantial information, simply report it truthfully. Your position will be that of a source of information, not an accuser. The matter will not be regarded as a complaint coming from you, but as a case that OIG or the institution is evaluating on its own behalf as a representative of the research community.

It should be noted that individuals may also elect to report alleged misconduct in research sponsored by PHS or NSF directly to the agencies. This is often done when a person is making allegations about misconduct involving someone at another institution. The agencies will normally conduct a confidential preliminary inquiry, contacting the accused for an explanation. If the agency is not satisfied with the response, it will usually defer the matter to the institution for further inquiry and investigation.

In some instances an inquiry into possible misconduct may be initiated without any complainant's allegation per se. For example, institutional officials may become aware of possible misconduct by chance in the course of routine administrative duties, or individuals may seek clarification as to the propriety of a colleague's behavior without submitting a formal complaint or allegation of misconduct.

Review of Allegations of Misconduct

Institutional procedures often provide for a preliminary and informal review to determine whether a formal inquiry and investigation are warranted. This review involves a checking of facts to determine if an allegation has any reasonable basis before the initiation of a formal process. This stage is not part of the federal requirements.

The initial stage of formal information-gathering and fact-finding is the *inquiry* to determine whether an allegation or apparent instance of misconduct has substance, and whether there is sufficient credible evidence to warrant a formal *investigation*. The inquiry is part of the federal requirements, and is intended to allow a careful look into the situation without tainting reputations of possibly innocent individuals. The institutional reviewing official may be a designated individual or panel. It is normal for the institution's reviewing official to provide the subject of the inquiry with a written statement or summary of allegations, and then invite response. In the rare circumstances where there is a good-faith concern that evidence will be destroyed or that a proper inquiry will be compromised, an institution may initiate certain investigative or protective actions before formal notification of the subject.

The reviewing official will consider the content and reliability of information received (e.g., degree of specificity, supporting documentation) and any prior knowledge of the individuals and events associated with the possible misconduct. Federal regulations attempt to assure peer review in the inquiry or investigation by requiring institutions to secure necessary and appropriate expertise to carry out a thorough and authoritative evaluation of the relevant evidence. The reviewing official will take appropriate steps to obtain or preserve all data (laboratory materials, specimens, and records) necessary to make a determination in the case. On completion of the initial inquiry, the reviewing official will prepare a written report that either dismisses the allegations or finds that an investigation of the allegations is warranted. This report will contain a statement of the evidence reviewed, summaries of relevant interviews, and the conclusions of the inquiry. A copy of this report will be provided to the individual against whom the allegation was made. This individual may submit comments to be made part of the record of the inquiry.

The initial inquiry should be carried out in a prompt manner; pursuant to the HHS/PHS regulations, it will normally be completed within sixty calendar days of initiation unless circumstances clearly warrant a longer period. NSF regulations allow ninety days for the completion of an inquiry.

A decision to dismiss the allegation is made when the reviewing official, after diligent and due investigative efforts, cannot uncover any reasonable substantiation for the allegation, or finds that the allegations as substantiated do not fall within the parameters of the federal or institutional definitions of misconduct. However, if the official finds any credible evidence that misconduct has occurred, the allegation will be advanced to the formal investigation stage.

If the reviewing official determines that the allegations should be dismissed, the corresponding report is maintained by the institution along with other sufficiently detailed documentation to permit a later assessment (if necessary) of the reasons for determining that an investigation was not warranted.

If, on the basis of the initial inquiry, the institution determines that an investigation is warranted, the institution's procedures will normally prescribe a formal process in which the person charged has certain *due process* protections. The greater the potential effect of the proceeding on an accused individual's liberty, reputation, or property interests, the greater must be the due process assured in a given situation. Although individual academic and research institutions are allowed considerable discretion in matters of governance, and enjoy local latitude as to the best means of implementing due process to suit their organizational culture and structure, the basic features of typical institutional procedures for formal investigation of research misconduct will normally include a *hearing* and extend the following rights to the accused: (1) to provide oral testimony to the investigating official (individual or panel); (2) to submit additional written statements and exhibits to the investigating official; (3) to submit questions for any witness to the investigating official for consideration; (4) to be accompanied by a lawyer or any other person when appearing before the investigating official (but, dependent on the particular institutional procedures, such person might be limited to serving as advisor only, and may not necessarily be permitted address or question witnesses or the investigating official or otherwise participate); and (5) to receive or inspect copies of all written materials presented to the investigating official. The individual charged may be required to provide the investigating official with such information relevant to the allegations as research notebooks, logs, source documents, lists of collaborators, lists of abstracts and papers, grant applications and reports, and other records. The NSF regulations allow 180 days, and the PHS regulations 120 days, for the completion of an investigation and a report of its evidence and conclusions.

In cases involving sponsored research funding, the awarding agency must be notified in accordance with the applicable regulations and requirements of that agency. PHS regulations require notification of the OSI when:

- on the basis of the initial inquiry, it is determined that an investigation is warranted;

- at any time the institution ascertains there is an immediate health hazard involved;
- there is an immediate need to protect federal funds or equipment;
- there is an immediate need to protect the interests of the person making allegations or of the person who is the subject of the allegations;
- if it is probable that the alleged incident is going to be reported publicly; or
- if there is a reasonable indication of possible criminal violation.

NSF regulations require prompt notification of the agency under similar circumstances.

Although NSF and PHS rely on research institutions to conduct inquiries and investigations, if they determine that the report of an institutional inquiry or investigation is not thorough, fair, objective, or responsive to regulatory requirements, the agencies may intervene and investigate allegations of misconduct directly. Also, in cases where misconduct is found, the agencies may impose additional sanctions, including written reprimand, institutional oversight, certification for future research applications, prohibition from service on government committees, or debarment. The agencies may initiate separate proceedings to adjudicate cases involving serious offenses for which severe sanctions are to be considered.

Protecting the Accuser and the Accused

Federal regulations require that institutional policies and procedures on research misconduct extend various protections to both the accuser and the accused. Federal and state law and institutional policies prohibit retaliation of any kind against a person who has reported or provided information about suspected or alleged misconduct, and who has not acted in bad faith. In this regard, it should be noted that if allegations are found to have been made in good faith, regardless of whether they were substantiated, no disciplinary or other legal action should be taken against the person who brought the charges, and warnings against retaliatory actions may be advisable. The PHS regulations require institutional policy to provide for the "undertaking of diligent efforts" to protect the positions and reputations of those persons who, in good faith, make allegations of research misconduct [42CFR 50.103 (d) (13)]. On the other hand, in the rare case where allegations of misconduct are found to have been maliciously motivated, appropriate disciplinary action or other legal measures may be taken against the accuser.

Federal regulations and typical institutional policies provide that privacy and confidentiality should be maintained to the maximum extent possible under relevant law. The confidentiality of the person bringing the charges should be protected as far as practicable to shield him or her from possible reprisals. Maintaining confidentiality of the identity of the accused individual until charges are initially substantiated will prevent needless reputational harm in situations where an allegation is the result of a misunderstanding or is otherwise without merit. However, whether a case can be reviewed effectively without the open involvement of the accuser and the accused depends on the nature of the allegation and the type of evidence available. If the case depends on the observations or statements of the accuser, it may not be possible to proceed without his or her open participation. In other cases, other non-testimonial evidence may allow a person making an allegation to remain anonymous.

A person accused of misconduct has an obligation to cooperate in good faith with the inquiry or investigation of possible misconduct. Refusal to cooperate and participate is a breach of that obligation, and may be dealt with as additional misconduct. Lying during interviews or other proceedings also may be considered misconduct and dealt with accordingly. A key question for persons accused of misconduct is whether to retain an attorney to represent them during an inquiry or investigation. Individuals are entitled to be advised and represented by counsel at their own expense at any stage of the proceedings, although the role of counsel may be limited at the inquiry and investigative stages as previously discussed. This decision is a personal one that should take into account the complexity and seriousness of the issues.

As discussed earlier, institutional policies must incorporate certain elements of due process to protect the accused in the course of a misconduct inquiry or investigation. Ideally, the process will also reflect sound peer review through the participation of objective fellow scientists and scholars as reviewers and expert advisors or witnesses. The presumption of innocence should prevail until a final determination regarding guilt is reached following the legally required procedures. A further safeguard for an accused individual is found in the *burden of proof*. The PHS and NSF regulations and typical institutional procedures use the "preponderance of evidence" standard; that is, a majority of the evidence (greater than 50 percent) must weigh in favor of the decision to find an individual culpable of misconduct in research. The PHS regulations further require institutional policies to assure precautions against real or apparent conflict of interest on the part of those involved in the inquiry or investigation. Finally, although the form and forum will vary according to state law and institutional policy, accused individuals must be afforded the opportunity for careful appellate review of such a decision that so profoundly affects their reputations and career interests. Accused individuals should also have their reputations protected by an adequate dissemination of exoneration when the process produces this finding.

Conclusion

The debate surrounding the adequacy of federal regulatory definitions and procedures regarding misconduct in research has not subsided. In May 1994, the Secretary of Health and Human Services appointed a commission to evaluate the Office of Research Integrity [(ORI), formerly, the Office of Scientific Integrity]. This commission has conducted hearings in which critics have complained about the perceived problems of a government office's making science policy through its decisions on whether particular conduct deviates from commonly accepted practices. Of further concern is the competence of reviewing officials to discern whether a researcher in a given case has committed culpable negligence, intentional misconduct, or excusable honest error. This continuing discussion may lead eventually to some refinement of the federal regulations.

Though the legal definitions and machinery surrounding the issue of misconduct in scientific research are appropriate safeguards to assure fair and effective handling of formal allegations and egregious conduct, main- tenance of the integrity of the research enterprise ultimately remains in the good hands of the scientific community itself. Scientists should be encour- aged and trusted to continue to exert stringent standards of conduct and accountability for each other, and to provide suitable training for new scientists in the values and methodologies essential to scientific inquiry. The scientific community's strong traditions of self-examination, disclosure, skepticism, experimentation/validation, and open criticism should expose the rare untrustworthy researcher and foster continuing public confidence in the scientific search for eternal truths.

Questions for Discussion

1. Assume that you and your advisor–professor have decided that you will pick up work on a project that was started and then abandoned a few years ago by a young faculty member on the research team. This faculty member authored a paper on his findings that was submitted but never accepted for publication in a scientific journal. After several weeks of

exhaustive attempts to replicate the experiment described in the faculty member's paper, you find that your results are markedly and consistently different (and less promising) from those previously reported by the faculty member. The faculty member has been evasive and obtuse when you have tried to discuss his procedures and data. In an effort to rule out the possibility of error, you confide your concerns to two fellow students and enlist their help in repeating the experiment, yet the group obtains results that match yours and not those reported by the faculty member.

a. Do you have a basis for suspecting the possibility of research misconduct on the part of the faculty member?

b. Do you have a responsibility to allege or report possible misconduct?

c. Should you go to your advisor–professor and/or confront the faculty member and/or take some other action?

d. Irrespective of what you do, do your fellow students have an independent responsibility to report this matter as possible research misconduct?

e. What should your advisor–professor do if you report your concerns to him or her?

f. What should you do if your advisor–professor's response to the situation is to sigh and advise that you should "forget about that project" and devise a new one?

g. If the faculty member is the subject of a misconduct inquiry and admits to "sloppiness" in his procedures, should the institution accept his confession and conclude the inquiry on that basis?

h. If the faculty member is found to have falsified the findings in his paper, can he nonetheless claim a defense or mitigating circumstance in the fact that his paper was never published, and he never otherwise reported the false data to the scientific community or the public?

2. Discuss the circumstances, if any, under which you believe negligent or "sloppy" research methods should be labelled and punished as misconduct. Discuss specific examples of practices and behaviors that, in your view, deviate from those commonly accepted in the scientific community.

   Does the prevalence of a shoddy laboratory technique weigh against a finding that it deviates from commonly accepted practices of the scientific community?

3. In your view, what are the elements of due process that should be afforded to the charged individual in the review of misconduct allega-

tions at the inquiry, investigation, and post-investigation stages, respectively?

Recommended Reading

Association of American Medical Colleges, *Beyond the "Framework": Institutional Considerations of Managing Allegations of Misconduct in Research.* Washington, D.C. 1992.

P.J. Friedman, *Research Ethics, Due Process, and Common Sense.* Journal of the American Medical Association 260:13 (1988), pp. 1937–1938.

B. Mishkin, *Responding to Scientific Misconduct: Due Process and Prevention.* Journal of the American Medical Association 260:13 (1988), pp. 1932–1936.

G. Taubes, *Bad Science: The Short Life and Weird Times of Cold Fusion.* Random, New York 1993.

Works Cited

42 CFR Part 50 (Code of Federal Regulations).

45 CFR Part 689 (Code of Federal Regulations).

Association of American Medical Colleges, *Beyond the "Framework": Institutional Considerations of Managing Allegations of Misconduct in Research.* Washington, D.C. 1992.

Black's Law Dictionary. West, St. Paul 1993, pp. 204, 226, 637.

P.J. Friedman, *Mistakes and Fraud in Medical Research.* Law, Medicine & Health Care 20.1–2 (1992), pp. 17–25.

National Academy of Sciences, National Academy of Engineering, Institute of Medicine, *Responsible Science: Ensuring the Integrity of the Research Process*, 2 vols. National Academy Press, Washington D.C. 1992.

Project on Scientific Fraud and Misconduct, *Deception in Scientific Research.* Report on Workshop Number One. American Association for the Advancement of Science, Washington, D.C. 1988.

A.H. Teich, M.S. Frankel (for the AAAS–ABA National Conference of Lawyers and Scientists), *Good Science and Responsible Scientists: Meeting the Challenge of Fraud and Misconduct in Science.* Directorate for Science and Policy Programs, AAAS, Washington D.C. 1992.

P.K. Woolf, *Science Needs Vigilance Not Vigilantes.* Journal of the American Medical Association 260.13 (1988), pp. 1939–1940.

CHAPTER TEN

Responsible Technology Transfer

Kenneth G. Preston

> The commercialization of molecular biology is the most
> stunning ethical event in the history of science.
>
> *Michael Crichton, Jurassic Park*

Introduction

Technology transfer has many definitions. To the teacher it may mean passing along to developing student minds an understanding of the physical relationships between rolling balls and inclined planes. To the global politician it may be finding ways to apply the results of modern science to the economic activity in developing countries (Samli). Technology transfer may be viewed broadly as the simple application of knowledge; alternatively from the perspective of applied science, it is the process of moving ideas from the research laboratory to the marketplace (Williams, Gibson, p. 10). For our purposes, we will consider technology transfer to be the process of identifying, evaluating, protecting, and commercializing the results of university research efforts, with special emphasis on interactions among and between university researchers and the private sector.

Ethics, too, may be described in many ways. Dictionaries use simplistic terms like "the discipline dealing with what is good and bad and with moral duty and obligation" (Webster's, p. 392) or "the branch of philosophy dealing with values relating to human conduct with respect to the rightness and wrongness of certain actions and to the goodness and badness of the motives and ends of such actions" (College Dictionary, p. 453). It may be useful to consider S. Gorovitz's view:

> Ethics is about interpersonal relationships, about how people and the institutions they constitute ought to behave toward one another. Ethics is therefore rooted in concern with peoples, individual and collective needs, aspirations, and values, and it requires a sensitive understanding of what those are. What is often overlooked is that ethics is autonomous. Ethical questions cannot be resolved simply on the basis of an appeal to the facts of psychology, economics, law, sociology or any other discipline. It is its own enterprise. Ethics is not a list of right actions and wrong actions, or a set of moral prescriptions and prohibitions. It is, instead, a discipline concerned with principles and with the quality of moral reasoning (Dinkel, Horisberger, Talo, p. 23).

It is in this context, then, that we approach the topic of responsible technology transfer. Our overall goal is to develop a discipline of moral reasoning to govern our personal interactions and decisions as we strive to find commercial applications for research results. More specifically, the following is intended to apply that reasoning discipline to the search for solutions to conflicts common to technology-transfer activities.

Information Management

The potential for conflict resides at every stage of the transfer process, and it is perceived, modified, and tempered generally by the motivations and obligations of the particular participants. Even at the very inception of research, the parties involved must exchange information that is often valuable and proprietary. The proposals and budgets that follow demonstrate quickly that the goals and objectives of researchers, universities, and sponsors can vary widely. And attendant to those varying objectives are divergent views as to measures of risk, success, and control—thus, conflict.

The mechanisms for exchange of confidential information have become routine for lawyers and administrators, and range from simple nondisclosure form letters to negotiated clauses in research and license contracts. The documents usually identify the subject matter in question and the purpose for the disclosures; establish the duration and extent of the nondisclosure obligations; and set forth exclusions to those obligations, such as cases in which the recipients already knew about the subject matter before disclosure, or the subject matter becomes public knowledge. Yet, while the parties may agree to these documents, the underlying goals of exclusivity, confidentiality, and the protection of trade secrets and know-how—so commonly believed necessary by the sponsor/commercializer—may be in opposition to researchers, desires to publish, exchange, teach, and seek knowledge. Areas of potential conflict include commercial success versus achievement, and wealth versus scientific acclaim. What makes it difficult, of course, is that all the goals may be worthy, just incongruous. The issues can become more difficult when someone begins to doubt the worthiness of a goal and questions whether it is right, or accepts the rightness of a goal but questions the means being employed to achieve it.

Most parties, whatever their roles may be in the technology-transfer process, avoid doing that which is clearly wrong. Indeed, many a decision is driven not by the decision-maker's view that it is right, but by a belief that the consequences of the decision will not be wrong. Thus, it may not be wrong to delay publication of research results until patent protection can be sought or until utility can be shown through experiment (Zurer "Scientists Confront", p. 21). It is not wrong, we may rationalize, because it is the only way for a commercializer to justify the risks in bringing products to the marketplace. And, after all, we may argue, what value is there to a technology that cannot be brought ultimately to the benefit of mankind? Yet such delays should be imposed only for reasonable times. The benefit of research results should not too long be withheld from the research community. Researchers' careers may depend on publishing; and often those participating in the research need to use the results for theses and dissertations. Thus, it may not be wrong, for example, to provide a sponsor several months in which to protect inventive ideas that may derive from a university's research for, in all practicality, it takes at least as long to publish a paper. Still, it may seem very wrong indeed to place the decision as to how and when, if ever, to publish a set of results entirely in the sponsor's hands. One suspects always the potential for corruption with such absolute control.

Intellectual Property Ownership

Addressed also at any stage, and sometimes at every stage of the transfer process, is the ownership of research results. Sponsors, with good reason, maintain that they have paid for the research by way of salaries for the researchers, equipment, supplies, and indirect costs to compensate the host universities. Thus, any results therefrom, inventive or otherwise, are derived under a "work for hire" principle that meets all the terms of an equitable definition, if not a legal one. Researchers may agree that information and data should be acquired by the sponsors, along with such researchers' reports, programs, and specimens as are clearly specified and required under the contract. Researchers understand that they will perform certain tests, record and analyze results, and then report same. But inventions are different. Inventions derive from the intellectuality of the researchers, and they can neither be predicted, defined, nor required in advance. Research contracts are not breached if researchers fail to invent. Thus, they will assert that inventions are, in reality, outside the research that was sponsored, and therefore subject to researchers' claims every bit as much as to those of the sponsor.

Universities are also participants in the ownership debate. Universities may well be in a position to argue that researchers are their employees, and that universities provide the resources, facilities, infrastructure, and environment necessary to the contract, having value well in excess of the indirect costs paid by sponsors. Universities, they will say, underwrite and subsidize entire research programs, and are entitled to the results. And so, each lays claim to the inventions and other intellectual properties. But for what purpose? The purpose, of course, is profit. And any reasonably-free-market, competitive, capitalistic system is driven by profit.

Private-sector corporate sponsors must be profitable to survive, and products and services supported by protected intellectual property may be key to their maintaining a competitive edge. Entrepreneurial researchers, after satisfying their professional interests and achievements in contributing to knowledge and advancing science, often seek to emulate, if only in a small way, the Microsofts, Genentecs, and other high-tech successes of recent decades. This entrepreneurial spirit is the backbone of our economic system. Universities, buffeted by rising costs, decreasing federal research dollars, and declining taxpayer support, consider participation in revenues from the commercialization of their technologies as viable means of supporting their research facilities and capabilities.

Thus, the interest by all parties in revenue participation is clear and, by most counts, meritorious. However, the pursuit of that interest, real or perceived, is the genesis of interest conflicts. The public is concerned, even cynical, about whether that pursuit will be carried forth in a manner consistent with the moral and ethical standards of the relevant constituent society and will appropriately safeguard public interests.

The issues of responsible technology transfer are not new. Landekich discusses many of the formal and informal private-sector codes of conduct and the responsibilities of management, employees, and internal audit (Landekich, pp. 2–9). Professions including law, medicine, and accounting have established rules of conduct by which practitioners of these professions are guided. Governments at all levels may establish codes, standards, statutes—even agencies—in an effort to safeguard the public interest. Public interest concerns may include a myriad of private and public-sector activities such as securities transactions, highway safety, environmental and consumer protection, food and drug manufacture and distribution, building construction, and antitrust and unfair business practices. Yet proposed rules and guidelines for the scientists involved have been marked by confusion in semantics, applications, and perceptions; and, as pointed out by Paul Friedman, they

> often demand the impossible, condemn the inevitable, and confuse poor judgement with dishonesty because their authors do not have a very precise idea of what constitutes a conflict of interest, what is bad about having a conflict of interest and, therefore, what are reasonable measures to take (Rachmeler).

Alberts and Shine comment that, historically, ethics was taught and learned informally among scientists, and that the scientific community was self-policing and self-correcting. They note the current heightened mistrust of scientists, and pleade for adherence by researchers to high ethical standards to avoid the constraints of "legal strictures, financial oversight, and bureaucratic provisions" being imposed upon them. They are, of course, correct in their assessment, as exemplified by the political pressure and threatened sanctions against Scripps Research Institute for its proposed technology-transfer agreement with Swiss-based Sandoz Pharmaceutical Corporation (Anderson). Under increasing pressure from both the public and private sectors, the Public Health Service and the National Science Foundation in 1995 established specific standards for professional conduct under federally sponsored research projects (Public Health Service; National Science Foundation). In essence, some of the common review and monitoring criteria that have long been accepted business practices in the financial world are now being applied to the scientific world, and in many respects, technology-transfer activity is being "audited."

Rules, standards, and laws in other nations may be more or less demanding than those of the United States, depending primarily on the culture of the people. International transactions raise different issues as a result of those cultural differences, coupled with differing local business practices and diverse political systems and involvements. For example, is it appropriate to transfer less than current technology to a less-developed nation? Politicians have criticized corporations for their imperialistic practice of selling out-moded, out-dated technology while maintaining their competitive edge with new developments. Realistically, the answer could depend on the needs of the people intending to use the technology, the availability of natural resources to support it, or whether the technology is being transferred for the economic purpose of creating jobs or the political–military purpose of developing weaponry (Weintraub, p. 26). We would expect a different set of answers if we asked which nation's standards should apply to the conduct of experiments and tests. The issues can be complex and vary greatly among nations, and cannot therefore be adequately dealt with here. However, successful transfer of technology among and between differing national cultures has been amply treated by others (Robinson; Kaynak; Stewart, Nihei).

Few of the foregoing standards of conduct are directed to public-sector corruption; rather, the focus is on controlling private-sector behavior within acceptable ethical boundaries while promoting achievement of the economic benefits of a free-market society. When that focus is directed specifically to technology transfer, questions about ownership again become paramount.

The U.S. federal government, the largest of the nation's research sponsors, determined in 1980 that it was in the best interest of the country, the economy, and the advancement of applied science for it to relinquish to contractors any claim of ownership to inventions made under federally sponsored contracts, retaining for itself only the traditional shopright—the right to make or use goods and services under such inventions for its own purposes (Public Law). This action stemmed from a recognition that in the free marketplace few inventions are deemed to be worth the risk of development expenditures without the promise of exclusivity. Hence, government ownership of government-sponsored research inventions thwarts the very goals of most federally sponsored research programs. With few exceptions, private commercialization of discoveries derived from basic research funded by the federal government does not seem to be a major ethical issue (Korenman S19). Many state, local municipality, and foundation sponsors adhere to the federal guidelines, especially since their needs are generally met under the shopright principle.

The goals and objectives of private-sector sponsors, on the other hand, often extend far beyond merely meeting their own needs for goods and services. Their purpose is to develop research results for the marketplace at large, and they often perceive the risks surrounding that development effort to be acceptable only under circumstances in which they have exclusive ownership of all protectable rights to the technology. It would make little sense to a corporate financial manager, for example, to risk the millions required to bring a new drug to the marketplace if that product could then be freely copied by a competitor who, unsaddled with such development costs, could afford to under-price the market.

Researchers and universities may likewise seek to maintain intellectual property ownership in order to attract investment capital, ensure royalty shares, or trade for commitments of research support. However, in most circumstances, actual ownership is irrelevant—only revenue sharing and exclusivity are of importance. Thus, the real need for resolution of the ownership issue should be sought not in the context of market risks, investments, and revenues, but in connection with managing conflicts of interest. Arthur Caplan, a noted ethicist, has written:

> Conflict of interest is the great challenge to administrators within universities who are trying to balance the mission of a university against the demands of industry and private-sector collaboration. It poses the greatest challenge to public trust and support for universities (Rachmeler).

Carl Djerassi states that patent and royalty policies are the origin of most conflicts of interest, and that, under one recommended policy, universities are assigned ownership of all intellectual property (Djerassi, p. 972). Experience with such a policy suggests that many of the financially driven conflicts identified by Djerassi are indeed minimized if not entirely resolved by that policy. Yet, perceptions to the contrary are commonplace.

People perceive, for example, that a sponsor may find it advantageous to suppress research results and maintain trade-secrets protection for competitive purposes. A corporate marketer might delay introduction of new products in order to fully depreciate company investments in earlier products, or to stage new-product introductions to stabilize the market and permit the management of obsolescence, replacement-parts inventories, and warranties to maximize profitability of the enterprise.

Likewise, a researcher–owner of technology may be viewed as risking a loss of objectivity in ongoing research, in the promotion of his or her developments, in peer review assignments, and in other institutional and sponsor responsibilities. In the worst case there is concern that the researcher's integrity may be compromised, bias may be introduced into the research results, students may be exploited in carrying out commercial research, or the direction of the research may be altered for personal gain; all of which can damage the public trust in the institution and the sponsor (Korenman S20).

The university-owner of a technology appears to be least likely to have real or perceived conflicts of interests whereby it would derive revenue from technology, primarily because the university neither performs research nor manufactures or markets products and services. In achieving the revenue objectives of furthering its teaching, research, and public-service goals, the university must rely wholly on the best efforts of the sponsor, the researcher, and, ultimately, the private-sector commercializer. Of the various potential technology owners, universities are least likely to be in a position or have the capacity to interject themselves into the technology development and utilization chain in a conflicting way. Nor do the objectives of researchers and private-sector developers seem in any way circumvented by university owners. Appropriate crafting of licenses can provide developers/marketers with all that is needed with respect to exclusivity of product lines and services, territories, sublicensing rights, and options for new developments, while researchers can share in the generated revenues and receive continued sponsorship for their laboratories.

New Ventures

An increasingly popular technology-transfer vehicle with respect to commercializing university research is the new-business-venture start-up. These businesses usually involve the researcher in some significant role, including one or more from among equity owner, consultant, scientific advisor, or corporate officer. The value of this kind of contribution to the success of the venture by a knowledgeable and zealous business and technology promoter or "champion" cannot be overstated, nor can the potential for real or perceived conflicts of interest and commitment. Many public and private-sector statements and policies regarding conflict of interest, including PHS and NSF guidelines, are directed to the kinds of conflicts most likely to occur in such start-up ventures. Often these policies are linked to scientific misconduct rules as well.

University owners of new-venture technology are in a unique position to manage and control researchers, conflicts, and to discourage misconduct. Appropriate exercise of that control can yield significant benefits to universities, researchers, and the public. Though start-up managers may maintain that company ownership of technology is critical to venture-capital fundraising, experience refutes this claim and demonstrates again that appropriate licenses under university-owned technology are effective. An additional benefit to university ownership of technology is linked to the high failure rate of start-ups; it is far easier to retrieve license rights from a failing or bankrupt start-up than to obtain reassignment of title-to-technology assets in the face of multiple creditor claimants. Thus, freeing the technology for further exploitation at another time is made far easier.

Even so, university ownership is no panacea for conflict issues. The avoidance of real and perceived conflicts of interest for the person attempting to be both researcher and start-up participant can be daunting, and often leads to a required choice between the two activities. Not surprising in such instances, a resolution of a *conflict in commitment*, that is, when outside activities interfere with fulfilling institutional responsibilities, is often every bit as critical as resolving *financial conflicts*. Susan Erhinghaus points out that most regulation is designed to address these two basic conflicts in which "an individual is potentially obligated or influenced to serve two masters so that his or her ability to act objectively appears to be or is compromised" (Erhinghaus, p. 2). Attempts to resolve such conflicts can lead to confrontations, and Erhinghaus succinctly summarizes the many legal challenges to the rights of governments and institutions to control researchers' activities.

Commercial Evaluations

Clinical trials and other evaluations of commercial marketability, especially those involving human subjects, have a high conflict-visibility with respect to institutions, the government, and the general public. In a typical case, little occurs in the way of technology transfer. The clinician receives a prototype pharmaceutical from the sponsoring company, administers it according to a pre-established test format, and reports the results. Hence, the concerns about conflict do not arise here out of technology development or transfer activities; rather, the concerns relate to the quality and integrity of the test results. Accordingly, the propriety of engaging testing clinicians with an equity or other financial interest in the sponsoring company or its technology is questioned and closely scrutinized.

Consulting

Consulting arrangements can confront researchers with many of the same issues and concerns as clinical trials. In addition, such arrangements can cause direct confrontations between a researcher's employing institution and the company for whom he or she is a consultant. Under common law the employer is entitled to at least a shopright in any inventions or other work products of its employees. Many universities and virtually all private-sector companies have extended that concept to include outright owner-ship of the employee's work product to the extent that it relates to the field or subject matter for which he or she was employed. On the other hand, many consulting agreements require the consultant to assign all rights in work under a contract to the sponsoring company, thus setting the stage for confrontation and dispute.

To avoid questions and misconceptions about the researcher's objectiv-ity, typical conflict policies preclude a researcher–consultant from being the principal investigator on research sponsored by the company for whom he or she consults, especially if the consulting duties include evaluation of the research results.

Small Business Innovation Research (SBIR) grants are treated in much the same way. Researchers who receive SBIR funding for companies in which they have an interest should not be the principal investigators for research subcontracted by those companies to the researchers, institutions. It is to be expected that SBIR grants will not be made in the absence of evidence that small-business entities will have access to technology developed thereunder. Thus, small-business entities, researchers, and institutions should define their interrelationships well in advance of anticipated grants.

Disclosure Requirements

In all of the foregoing, disclosure is the cornerstone of policies, standards, and regulations designed to avoid or control conflicts. Institutional rules often require annual or other systematic disclosure of all researchers, activities outside those assigned within that institution. Others, like the PHS and NSF guidelines, require full disclosure of such related activities during the sponsored-research granting process. Congressman Ron Wyden (D.-Ore.), a sharp critic of PHS's lack of action in the past to monitor and control public–private technology activities, believes that "a good part of the issue can be dealt with through disclosure . . . we need to make sure the public can get access to information about such arrangements" (Zurer "Latest Conflict", p. 21). There is a growing consensus that disclosure of researcher activities attended by first-line supervisory oversight and control is the most effective and least intrusive of the various proposals for avoiding conflicts or, if unavoidable, then for controlling them and promoting responsible technology transfer (Riley). Lest anyone surmise to the contrary, it is indeed the promotion of and encouragement for such transfer that is clearly being sought by all parties—research institutions, researchers, governments, the private sector, and the public.

CASE STUDY 10 A

Ethics and the Dispensing of Prescription Drugs

Should medical ethics or business ethics apply to the dispensing of prescription drugs? The *New York Times* has reported a rash of mergers among drug manufacturers and pharmaceutical service companies. For example, Merck acquired Medco, a company that supplies drugs through pharmacies and mail orders to thirty-eight million Americans; Smith Kline Beecham bought Diversified Pharmaceutical Services, which handles prescription drugs for eleven million customers; and Eli Lilly announced the purchase of PCS Health Systems, which has fifty million subscribers. At least in the Merck–Medco arrangement, Medco pharmacists are paid cash commissions for switching prescriptions from the products of other companies to those of Merck. Upjohn Company and Miles Inc. have agreed to pay fines in settlement of disputes in a number of states for paying pharmacists to steer customers to their products. On the other hand, Dr. Alan Hillman, Director of the Center for Health Policy at the University of Pennsylvania, finds "nothing illegal or unethical about it. It's just American companies trying to get a leg up in the marketplace." Under what circumstances should these business arrangements be considered unethical? How does this differ from doctors who send their patients for lab tests to diagnostic laboratories they own?

What are the issues to be resolved?

Source: Associated Press. "Drug Firms Dangle Offers." *New York Times* 31 July 1-9H.

CASE STUDY 10 B

Ownership and Conflict of Interest

Professor McCloskey in the Environmental Engineering Department is the university expert on water treatment. He had a consulting contract with Apex Inc. that required him to assign inventions to Apex. In the course of the contract, McCloskey developed a patentable sewage-treatment control system for Apex. It is believed that the invention was made during the summer with the use, to some extent, of university lab equipment.

McCloskey also invented a method of removing heavy metals from sludge. He claimed the invention for himself because he was a part-time employee, the invention was outside his field of expertise, and the invention was made off-campus.

One of McCloskey's graduate students, Gerry Smolenski, worked as a laboratory assistant under a university research contract sponsored by Baker Corporation, according to which Baker was to receive and own the research results. Smolenski claimed the research results were his own ideas, and then took the lab notebooks, patented the inventions, and sold the patent rights to a third party.

Another of McCloskey's students, Bill Reece, invented a flow meter during the course of his master's thesis research. McCloskey claimed to be a joint inventor, and insisted on being joint author of a paper to be published. Reece said McCloskey invented nothing, but threatened to approve the thesis only if Reece assigned him part ownership of the invention.

Reece later obtained employment as a researcher for Apex. After two years, he and another employee left Apex and started a company to manufacture control systems that are sold, in some circumstances, in competition with Apex. Apex sued for theft of trade secrets, unfair competition, breach of implied employment contract, and patent infringement.

What are the issues to be resolved?

Recommended Reading

For emphasis on technology transfer, see

R.D. Robinson, *The International Transfer of Technology*. Ballinger, Cambridge 1988.

A.C. Samli, *Technology Transfer*. Quorum, Westport 1985.

F. Williams, D.V. Gibson, *Technology Transfer*. Sage, Newbury Park 1990.

For emphasis on conflicts, see:

R. Dinkel, B. Horisberger, K.W. Talo, *Compromising Drug Safety: A Joint Responsibility*. Springer, New York 1991.

S. Landekich, *Corporate Codes of Conduct*. National Association of Accountants, Montvale 1989.

Works Cited

B. Alberts, K. Shine. *Scientists and the Integrity of Research*. Science 266 (1994), pp. 1660–61.

C. Anderson, *Scripps to Get Less from Sandoz*. Science 264 (1994), p. 1077.

College Dictionary. Random, New York 1982.

R. Dinkel, B. Horisberger, K.W. Talo, *Compromising Drug Safety: A Joint Responsibility*. Springer, New York 1991.

C. Djerassi, *Investigator Financial Disclosure Policy*. Science 261 (1993).

S.H. Erhinghaus, *Conflicts of Interest and Technology Transfer*. University of North Carolina, Chapel Hill 1993.

E. Kaynak, *Sociopolitical Aspects of International Marketing*. Haworth Press, Binghamton 1991.

S.G. Korenman, *Conflicts of Interest and Commercialization of Research*. Academic Medicine 68 (1993).

S. Landekich, *Corporate Codes of Conduct*. National Association of Accountants, Montvale 1989.

National Science Foundation, *Investigator Financial Disclosure Policy*. Federal Register 59:123 (1994), 33308–33312.

Public Health Service, *Objectivity in Research*. Federal Register 59:123 (1994), 33242–33251.

Public Law 96-517, *The Bayh–Dole Act*. Patent and Trademark Act Amendments of 1980.

M. Rachmeler, *Conflicts of Interest and Technology Transfer*. AUTM Manual (1994) vol. 3, II-2-2.

S. Riley, *Disclosure to Whom?* The News and Observer, 3 Apr. 1994, 13A.

Robinson, Richard D, *The International Transfer of Technology*. Ballinger, Cambridge 1988.

A.C. Samli, *Technology Transfer*. Quorum, Westport 1985.

C.T. Stewart, Jr., Y. Nihei, *Technology Transfer and Human Factors*. Lexington Books, Lexington 1987.

Webster's New Collegiate Dictionary. Merriam-Webster, Springfield 1973.

S. Weintraub, Sidney, *Technology Transfer and Soft Technology Development*. National Forum, the Phi Kappa Phi Journal LXI.1 (Winter 1981), pp. 25–26.

F. Williams, D.V. Gibson, *Technology Transfer*. Sage, Newbury Park 1990.

P.S. Zurer, *Latest Conflict-of-Interest Rules Mollify Critics*. C&EN 18 Oct 1993.

P.S. Zurer, *Scientists Confront Ethical Challenges*. C&EN 14 Mar. 1994.

Glossary:
The Language of Ethics

a posteriori **reasoning**: From effect to cause; making decisions based on the gathering of knowledge gained from experience.

a priori **reasoning**: Statements we can know to be true prior to any examination of the facts of the world; from cause to effect; making determinations based on what is understood to be universally true.

accountability: The quality or state of being answerable, liable, or responsible.

applied ethics: The application of moral standards in decision-making related to concrete rather than abstract conditions.

autonomy: The power to make moral choices and be self-governing; an ethical principle based on the union of nationality and freedom.

beneficence: An ethical principle of helping others whenever possible; active goodness, kindness, generosity; the quality of doing good.

bioethics: The discipline of assessing the rightness or wrongness of acts performed within the life sciences.

categorical imperative: A moral obligation or command that is unconditionally and universally binding.

conflict of interest: The predicament arising when a person confronts two actions that cannot be ethically reconciled; competing loyalties and concerns regarding self-dealing, outside compensation; divided loyalties among, for example, public and/or professional duties and private or personal affairs.

consequentialist reasoning: Reasoning in which the rightness of an act is linked with the goodness of the state of affairs that it brings about.

deontological reasoning: A type of reasoning which focuses not on the consequences of an action, but on the theory or study of duty or moral obligation.

deontologism: A moral philosophy in which acts are based on a self-determined, inner sense of moral "duty." Someone who adopts the examples of the divine command theory, for example, adopts the will of God as "the inner sense of duty." But a deontologist may choose any set of values as the basis for this innate comprehension of duty.

descriptive ethics: A theory which holds that only descriptive or empirical statements are meaningful; a branch of ethics that discusses the moral and ethical beliefs and customs of the peoples of the world.

distributive justice: The ethical principle concerned with the fair allocation of privileges, duties, and goods within a society in accordance with merit, need, work, or other agreed upon criteria.

due process: Specific and systematic procedures to which workers are entitled for appealing disciplinary and discharge action by employers.

duty: An action required by one's position or by moral or legal consideration, often contrasted with personal inclination or pleasure.

empirical: Referring to the gathering of information to verify statements.

empirical statements: Actual assertions about the world and our physical environment, verifiable by controlled observation, experiment, or direct sensory experience: usually contrasted with theoretical statements.

epistemology: A field of philosophy concerned with the theory of knowledge.

ethics: A discipline related to what is good and badincluding moral duty and obligation, values, and beliefsused in critical thinking about human problems.

fiduciary: Involving confidence or trust; a fiduciary obligation is one arising out of trust.

inductive reasoning: Thinking or reasoning from within, and applying the conclusion generally; usually based on intuitively measuring the probability of its application.

inference: Statement that can be drawn from combining propositions and variables; contrasted with "implication," which is the intended suggestion of the maker of the statement.

informed consent: A legal term referring to the process by which a person who, impression of suitable information, grants authority to someone else to take actions affecting that person; in medical ethics it indicates the patient's approval of a procedure or treatment, based on possession and understanding of all relevant information.

justice: Fair dealing or right action; the principle that demands that we subject our actions to rules, and that the rules be the same for all.

lex talionis: The law of retaliation, which says that no vengeance shall exceed the original hurt: an eye for an eye, a tooth for a tooth.

libertarian: A theory which gives priority to the principles of individual liberty and freedom of thought and of action, no matter how unwise a choice may seem to others.

logic: A system of evaluating statements or arguments.

maxim: A general truth, basic principle, or rule of conduct; often expressed as a proverb or saying.

mean: Moderate actions or attitudes appropriately chosen between two extremes; identified with virtue.

mens rea: "Guilty mind" or intent to injure; the assignment of lesser penalties for lesser degrees of guilt.

metaethics: Theoretical ethics, a discipline that considers the foundation of ethics, specifically the meaning of ethical terms and forms of ethical argument.

metaphysics: A branch of philosophy concerned with questions about reality and the nature of existence.

moral: Capable of distinguishing right from wrong with a predilection for right; as an adjective, it describes a person or act or thing that conforms to an agreed-upon standard of conduct; as a noun, it is a summation of truth from an incident or parable.

moral agency: The ability to make ethical choices and take responsibility for those choices.

moral development: Human growth in awareness of rightness or wrongness of actions, often accompanying maturity.

moral reasoning: Any process that applies general moral principles to a situation or human problem and reaches a conclusion or decision.

moral judgement: A personal conclusion about the rightness or wrongness of an action in a particular set of circumstances based on general principles.

moral law: A rule or group of rules of right living conceived as universal and unchanging.

morality: The inherent right or wrong nature of an action or conduct.

motive: The basis, reasoned or irrational, for the way a person acts.

non-maleficence: Avoiding harm or evil.

normative ethics: A discipline dealing with the nature of ethical principles that have been accepted as norms of right behavior.

obligations: Duties or debts to others; constraints on behavior because of a favor granted by or received from another.

philosophy: The study of principles that underlie human conduct and order in the universe; the study of reality.

plagiarism: An action whereby one appropriates another's writings or works of art and makes use of them as one's own; implied is the notion that by so doing, one is "stealing" original work produced by another.

prima facie: From Latin, "at first sight," or having an obvious meaning adequate to establish fact unless refuted.

relativism: Characterized by the notion that no point of view is more correct than any other; in matters of policy and ethics, the belief that there is only subjective (personal) opinion, and no objective (interpersonal) truth.

retributive justice: In ethics, the principle of justice concerned with punishing an individual for his or her actions.

Socratic method: The method of questioning that Socrates used to approach truth, which is assumed to be implicitly known to all rational beings.

sophists: Literally "wise ones"; these were the professional teachers of rhetoric in ancient Greece, who taught that the selfish life was the best life.

stoicism: A philosophical system of the Stoics, who held that our duty is to conform to natural law and accept our destinies, and that wise human beings should remain unswerved by joy or grief.

truth: Objectively accurate statements, always applicable.

utilitarianism: A subset of consequentialism in which the guiding principle is to always act in such way as to produce the most pleasure, fulfillment, or happiness for the greatest number of people; prime advocates include Jeremy Bentham and John Stuart Mill.

value judgement: A judgement that assigns good or evil to an action or entity.

virtue: A single term conveying all the qualities comprising excellence in human beings.

whistle-blowing: Bringing into public view an employer's neglectful or abusive practices that threaten the public interest.

Index

H.H. Sedlacek/A.M. Sapienza/V. Eid

Ways to Successful Strategies in Drug Research and Development

1996. XIV, 260 pages. Hardcover. DM 128,-. ISBN 3-527-29415-5

Strategic planning is a critical subject, central to the success of any scientific and economical enterprise. Not only is the scientific knowledge of many persons needed, but also an assessment of what may occur in the future - which approach may be competitive, which option can be achieved, and how can this be accomplished.

With a focus on the various ethical obligations to patients, animals and the environment, this book offers hands-on help on how to develop successful R&D strategies, taking special account of the needs of scientists and managers in the pharmaceutical industry. Key topics include:

- evaluation and selection of projects
- measures to reduce risks
- project management
- corporate and technology strategy
- managing for innovation

The reader will learn the methods needed to elaborate strategies so that he or she will become aware of the numerous managerial, organizational, social and political parameters and forces, the consideration of which is essential for the successful realization of a formulated strategy.